WINDBORNE PESTS AND DISEASES:
Meteorology of Airborne Organisms

WINDBORNE PESTS
AND DISEASES
Meteorology of Airborne Organisms

DAVID E. PEDGLEY
Research Meteorologist
Centre for Overseas Pest Research, London

ELLIS HORWOOD LIMITED
Publishers · Chichester

Halsted Press: a division of
JOHN WILEY & SONS
New York · Brisbane · Chichester · Toronto

First published in 1982 by
ELLIS HORWOOD LIMITED
Market Cross House, Cooper Street, Chichester, West Sussex, PO19 1EB, England

The publisher's colophon is reproduced from James Gillison's drawing of the ancient Market Cross, Chichester.

Distributors:

Australia, New Zealand, South-east Asia:
Jacaranda-Wiley Ltd., Jacaranda Press,
JOHN WILEY & SONS INC.,
G.P.O. Box 859, Brisbane, Queensland 40001, Australia

Canada:
JOHN WILEY & SONS CANADA LIMITED
22 Worcester Road, Rexdale, Ontario, Canada.

Europe, Africa:
JOHN WILEY & SONS LIMITED
Baffins Lane, Chichester, West Sussex, England.

North and South America and the rest of the world:
Halsted Press: a division of
JOHN WILEY & SONS
605 Third Avenue, New York, N.Y. 10016, U.S.A.

© 1982 D. E. Pedgley/Ellis Horwood Ltd.

British Library Cataloguing in Publication Data
Pedgley, David E.
Windborne pests and diseases: Meteorology of Airborne Organisms
1. Pests
I. Title
632   SB601

Library of Congress Card No. 82-9197 AACR2

ISBN 0-85312-312-8 (Ellis Horwood Limited)
ISBN 0-470-27516-2 (Halsted Press)

Typeset in Press Roman by Ellis Horwood Ltd.
Printed in Great Britain by R. J. Acford, Chichester.

Table of Contents

Foreword

FEATURES OF THE ATMOSPHERIC ENVIRONMENT

By R. C. RAINEY, O.B.E., D.Sc., A.R.C.Sc., F.R.S.
former President, Royal Entomological Society, London

To the insects, birds and other living beings which the air carries (and to the meteorologist), the atmosphere is by no means the uniform and structureless medium which it may appear to the earthbound biologist. Atmospheric structures and organised systems range in scale from the trade-winds and monsoons of the global circulation, and the depressions and anticyclones of the weather forecaster, to the thermal up-currents sought by the glider pilot, and the minute eddies which disperse the invisible plume of airborne sex-attractant from the female moth.

Many of these atmospheric structures and systems have been found to be of major importance in the life of airborne organisms, and in turn significant in dealing with problems of airborne pests and diseases. Effects of winds and weather are spectacularly manifested by the desert locust, and so it was, twenty years ago, that the Desert Locust Control Organization for Eastern Africa, newly established and maintained by the governments of Ethiopia, Kenya, Somalia, Tanzania, Uganda and the then French Somaliland, approached our Meteorological Office for the secondment of a meteorologist of appropriate interests and experience.

The immediate problem was the wind-systems over and around the Red Sea, particularly the zone of wind-convergence which is regularly to be found there during the winter months and, although very little studied meteorologically, was already known to be much involved in the complicated patterns of migration and breeding of desert locusts in this region. The meteorologist was David Pedgley; he had first shown his flair for elucidating such atmospheric features when as a meteorological observer in the then Suez Canal Zone he had undertaken (initially just as a hobby) supplementary pilot-balloon observations which had revealed the structures of the Sinai sea-breezes. Back in England, he had demonstrated such pilot-balloon observations of upper winds at field study courses on weather and bird migration, which had introduced him not only to the biological significance of wind-systems but also to Professor R. S. Scorer, now editor of the present Environmental Sciences series – as well as to the writer of this foreword.

A preliminary report on David Pedgley's studies of the Red Sea convergence zone, utilising the facilities and resources of the locust control organization, was published in *Weather* in 1966, and earned him the Darton Prize of the Royal Meteorological Society. Since then, scope and material for the application and extension of these findings has been provided by two desert locust upsurges in this region, as well as by devastating attacks of another airborne migrant, the African armyworm, on the crops of the Yemen Arab Republic – the ancient Arabia Felix.

This book provides a fascinating introduction to the widely-scattered and rapidly expanding literature on airborne pests and diseases, including striking recent radar findings, and with basic explanations of many of the aspects of meteorology involved. It reviews extensive evidence of a dominating role of wind-systems in the movements and distribution of many organisms, and will encourage the reader to consult original papers for further details and implications.

R. C. Rainey
June 1982   Stoke Mandeville

Preface

This book describes and explains the influence of the atmosphere on the windborne movement of small organisms, and how they get into and out of the atmosphere. It is a book for biologists by a meteorologist. In it I have attempted to increase interdisciplinary understanding of complex practical problems in pest and disease management, and to highlight the need for multidisciplinary work on solutions to those problems. Biologists, both student and practising, who are involved with organisms that get into and are then moved by the air, include those concerned with not only pests and diseases but also ecology more generally. Others with related interests are naturalists and agricultural meteorologists.

Many biologists have had little or no formal instruction in meteorology. It is therefore perhaps not surprising that they tend to ignore the possibilities and consequences of windborne movement, even though examples are familiar to us all: the sudden arrival of aphids and locusts; the development of hay fever and mouldy food. Nevertheless, there is no doubt that the application of meteorology to aerobiology can lead to improvements in forecasting, monitoring and managing windborne pests and diseases.

Biologist readers must not expect too much biology from a meteorologist author. My aim is not to give a catalogue of known windborne pests and diseases, or of established management practices, nor to discuss the significance of adaptation to windborne movement, but to explain general principles and to encourage their wider application by means of illustrations taken from the literature. Some of these examples are treated intentionally very briefly, but they have been included to help strengthen particular lines of argument. The reader will be able to add his own knowledge and experience. Some examples are not concerned with pests and diseseases, but they complement the discussion, which is based largely on studies of harmful organisms, for it is upon those organisms that most resources have been spent.

Only those aspects of meteorology have been introduced that seem to be directly helpful; the reader will be able to find much more in standard texts, some of which are referred to. Chapters 1 and 2 deal with atmospheric structure and behaviour, and their effects on take-off, survival and landing, but not with development of readiness to take off or changes after landing – they are subjects too large for this book. Chapters 3-7 are more concerned with weather as commonly understood – the causes of day-to-day changes in wind, sun and rain, and how they bring about movements of organisms, sometimes over hundreds or thousands of kilometres. Chapters 4 and 5 deal with the flight of insects as individuals, and Chapter 6 with swarms. Chapter 7 is concerned with the ways in which weather affects the spreading apart and gathering together of clouds of organisms. Lastly, Chapter 8 discusses and illustrates some of the requirements of services needed to forecast and manage the windborne spread of harmful organisms. These chapters vary greatly in length, but they reflect the differences in effort that have been put into research.

Much has yet to be done to understand and to reduce the spread of windborne pests and diseases. Much research will be needed on the effects of weather on not only the mechanisms of take-off, survival and landing, of spread, and of dispersion whilst airborne, but also their day-to-day application to whole populations of organisms in the field. Operationally, much will need to be done to improve both field monitoring techniques, and the analysis of records, to make possible the provision of timely warnings and forecasts. The challenges are great. Effective management of pests and diseases is often beyond the means of the individual citizen; it will require the cooperation of many disciplines.

Some aspects of atmospheric influences on airborne organisms have been discussed in other books, notably Johnson (1969) and Gregory (1973). It is a pleasure to acknowledge the inspiration I have received from these two authors, both from their writings and in discussion. My colleagues at the Centre for Overseas Pest Research have also contributed to this book, both directly and indirectly, and I thank particularly Dr R F Chapman for reading the manuscript and suggesting improvements. Among others who have helped me are Professor C T Ingold and Professor R P Pearce. I thank my wife for drawing the diagrams; many of them have been adapted from the literature, but in all cases the original authors have been acknowledged.

February 1982   D.E.P.

CHAPTER 1

Weather at take-off

Windborne tufts of seed-carrying down from thistle (*Cirsium*), willowherb
(*Epilobium*) and other plants show clearly how some organisms have adapted to
using the wind as an aid in seeking new habitats in which to continue the species.
Mass take-off by some insect species provides other examples. But less spectacular,
and indeed often unnoticed, is the take-off and windborne spread of many
other kinds of organisms — often very small, like pollen grains and fungal spores.
Such take-off is often thought to be of two contrasting kinds: active and passive.
Active take-off involves the use of energy from *within* the organism — such as a
spring- or gun-like mechanism that shoots out pollen grains or spores, after
being primed by changes in temperature or air humidity; or the jumping or
flying by insects from a resting place, in response to some trigger such as a
change in temperature or wind. *Passive* take-off involves the use of energy from
outside the organism — for example, the energy of a wind gust or a falling rain
drop that can knock an organism into the air. But this distinction between active
and passive is not always clear, for the ability of a gust to knock, say, the seeds
from a plant may well depend upon that plant having the ability to produce
seeds that can be so knocked.

These examples show it is weather *changes* that can launch organisms
into the air. We need therefore to know and understand these changes. It is
convenient to separate two time scales: diurnal (day-to-night) and sudden
(within minutes or seconds). They are discussed in sections 1.3 and 1.4, but
first we look at some of the observed effects of weather changes on take-off.

1.1 SOME EXAMPLES

In this section we consider some examples of how organisms become airborne
as a result of weather changes. The examples are merely illustrations, and the
reader can no doubt add others from his own experience. For simplicity, the
weather parameters of temperature, humidity, wind and rain are treated separately, but it should be remembered that their effects are sometimes combined

and therefore difficult, or as yet impossible, to disentangle when two or more parameters change together. For more extensive discussion of the mechanisms of take-off, the reader is referred to standard works, such as Ingold (1971), Pijl (1972), Proctor & Yeo (1973) and Hers & Winkler (1973).

1.1.1 Temperature

The crackling of gorse (*Ulex*) on a warm, sunny day is well known in some countries; the seed pods suddenly snap open by twisting, thereby throwing their seeds into the air. This active take-off comes about as the two halves of a pod turn into springs when they dry out, and tension builds up until it overcomes the bond between them. Drying out starts when the water supply to the pod becomes restricted, but it is no doubt also affected by progressive changes in both temperature and air humidity. In sunshine, the seed pod is warmer than the air (page 32), and evaporation of tissue water is therefore encouraged. Moreover, as the day becomes warmer so the relative humidity (page 35) of the air becomes less, thereby further encouraging drying out. These **seeds**, shed explosively (as in other species, such as some in the families Geraniaceae and Euphorbiaceae), are perhaps hardly airborne, but their release mechanism emphasises the need to consider atmospheric changes right next to the organism concerned, and not at some convenient place more or less nearby. More clearly airborne is **pollen** (of some flowering plants − for example, of the family Urticaceae) which is shot into the air by the sudden unbending of strained anthers, very likely influenced by temperature and humidity changes.

Some small, **wingless insects** get into the air by releasing their hold on the place where they had been standing, and then falling with or without the help of silk threads (see pages 89 and 91). The same is true of some arachnids such as spiders and mites. Like all activities of animals, this take-off is likely to be temperature dependent, and there may well be **temperature thresholds** for each species, below which active take-off never takes place. Diurnal temperature changes (page 42) may therefore affect the timing of such take-off. Some **scale insects** (order Homoptera) show this effect, and so do the leaf-feeding **caterpillars** of some moths.

Another animal activity that can put organisms into the air is breathing, especially the heavy breathing of mammals with infections of their respiratory systems. Infected mucus can be breathed out as droplets, particularly during coughing and sneezing − both of which can be brought about by changes of air temperature.

Winged insects take off by flapping flight, but again only if the temperature is above a threshold that varies with species, and to some extent between individuals of the same species. Even then, take-off usually needs to be triggered. It is the temperature of the thoracic flight muscles that is relevant, for the muscles must be warm enough if they are to give the power needed for take-off. A large insect basking in sunshine can be 10°C or more warmer than the air around it. Even a small insect, such as an aphid or a thrips, can be as warm as the plant or ground it stands on, when being warmed by the sun. Because temperature can vary by several degrees between sunny and shaded parts of a plant, even across a single leaf, the take-off threshold for a particular species can be exceeded in some parts and not in others. Insects sometimes crawl to sunny spots so that by basking they become warm enough for take-off. On the other hand, the passing of a cloud across the sun can cause leaf temperature to fall enough to stop take-off.

Take-off rate has been well studied for the **bean aphid,** *Aphis fabae,* in summer (Johnson & Taylor 1957), but many other insects may well take off in much the same way. On fine, warm days, flying density above a settled crowd of this species has two peaks: early morning and late afternoon. Newly fledged insects need some heat (expressed as degree-hours above a threshold) before flight can be triggered. Because temperature varies diurnally (page 42), the time needed to get this heat also varies with the time of day. Those that become ready to fly during the night do not do so because of cold and darkness, so take-off is held back until the temperature threshold for flight is passed, thereby leading to the early morning peak. During the day, as temperature rises, the rate at which aphids become ready to fly quickly grows, so the volume density of insects flying just above the source reaches another peak, after which the density falls away as darkness comes on. A smaller density around midday can be because too few aphids are ready to take off, and not because the weather is against flight or because insects are carried aloft and out of sight. If the night is cold, too few aphids are ready by the time the flight threshold is reached, and so the morning peak is lost. If the day is too warm, the rate at which aphids become ready to fly is also slowed down and so the afternoon peak may be lost. With *Aphis fabae,* the range of temperatures for take-off is 9 to 28°C; it will not be the same for other species.

1.1.2 Humidity

The effects of humidity changes on take-off rate of insects has been little studied, but laboratory experiments with viviparous females of the **cereal aphids** *Sitobion avenae* and *Rhopalosiphum padi* showed that a fall in *relative* humidity over a few minutes increased the take-off rate, whereas a rise decreased it (Rautapää 1979, Fig. 1). Similar results were obtained with the **coccinellid** *Myrrha 18-guttata* (Pulliainen 1964). Humidity was controlled by passing the air over water or over saturated solutions of various salts.

Active **spore** release is known to be affected by changes in relative humidity (page 35). Many species of **fungi, mosses, leafy liverworts** and **ferns** actively throw their spores into the air, some as far as several centimetres. Some fungi put forth their spores in damp air, perhaps because a high relative humidity is needed for a swelling up of the spore-bearing tissues, which then bend or burst

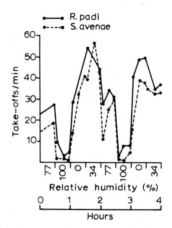

Fig. 1 – Effect of alternating relative humidity on take-off rate by female cereal aphids *Rhopalosiphum padi* and *Sitobion avenae*. Relative humidity was changed every half hour and take-offs were counted 5, 15 and 25 min after each change. Means for eight groups of 40 aphids. (After Rautapää 1979).

suddenly and so shoot off the spores. For example, many of the fungi **Asco-mycetes** have fruiting bodies in the form of tiny sealed bags (asci) filled with sap in which the spores (ascospores) are lying. When the inside pressure becomes great enough, the bag opens suddenly and the spore-laden sap squirts out. This squirt-gun mechanism can be triggered also by a *fall* in relative humidity. In laboratory experiments with *Calonectria crotalariae*, a fungus that causes severe **black rot disease of peanuts** and other legumes, Rowe & Beute (1975) found that countless individual ascospores were shot off when the relative humidity fell from 100% to 75% within about 15 min (Fig. 2). After returning to 100%, a similar fall gave another massive discharge. With *Sordaria fimicola,* similar experiments starting at 95% relative humidity showed that spore discharge with a second fall in relative humidity was smaller for larger falls (Austin 1968),

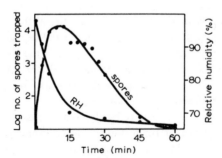

Fig. 2   Effect of falling relative humidity on release of ascospores of the black rot fungus of peanuts, *Calonectria crotalariae*. (After Rose & Beute 1975).

probably because the number of spores in the first discharge was greater for larger falls, and there was not enough time in the experiments for the development of many new asci (Fig. 3). Sudden drying gave maximum discharge rate after about one minute. Ascospores of **ergot**, *Claviceps purpurea*, are shot off during sunshine after a shower, probably as relative humidity falls.

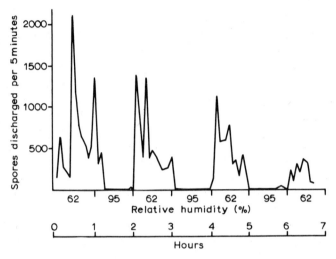

Fig. 3 – Effect of alternating relative humidity on release rate of spores of the fungus *Sordaria fimicola*. Culture grown at 95% relative humidity for 15 h before the hourly changes started and spores counted every 5 min. Air flow at 200 ml/min. (After Austin 1968).

By contrast, the fungi **Basidiomycetes** have their spores not within bags but at the tips of fine threads (basidia). A spore is shot off just after a drop, or a bubble, suddenly grows at its base, where it is fixed. Beneath the well-known cap-shaped fruiting bodies of mushrooms and toadstools (raised from the ground on stiff stalks) and bracket fungi (growing from the sides of trees) there are countless basidia from which spores are shed and fall into the air beneath. Bohaychuck & Whitney (1973) examined the shedding of basidio-spores of *Polyporus tomentosus*, a cause of **stand-opening disease in white spruce**, *Picea glauca*, in parts of Canada. Shedding rate in the laboratory (where temperature, humidity and light could be varied at will) was found to be greatest in darkness with a relative humidity greater than 85%, and the rate increased as temperature rose from 5° to 23°C. If relative humidity fell below 70%, far fewer spores were shed. In the field, shedding rate was greatest on mild, moist, cloudy days, especially in the evening, when the relative humidity was rising (page 44).

Other kinds of fungi put forth their spores when the relative humidity falls. In the small, cup-shaped fruiting bodies (aescia) of rusts, the walls of cells next to the spores become strained and mishapen as they dry out, but they then suddenly round off, pushing spores into the air. Some of the fungi **Phyco-mycetes** throw off their spores as the branched threads on which they have grown begin to dry out and twist jerkily after sunrise. These movements were studied in the laboratory with *Peronospora trifoliorum*, the cause of **downy mildew in alfalfa** (lucerne), *Medicago sativa*, after being grown at 100% relative humidity and then suddenly exposed to dry air (Fried & Stuteville 1977). The sporangia (parts where the spores were formed) shrivelled within seconds, and the twisting of the sporangiophores (stalks bearing the sporangia) continued for up to a minute, with discharge of some sporangia. A similar spring-like launching mechanism shoots the **mite** *Rhynacus breitlowi* from the underside of leaves of *Magnolia grandiflora,* where it lives (Davis 1964). In nearly saturated air, the hairs on the leaves become matted, but on drying out a little they spring upright, and mites are thrown as much as 10 cm. Other laboratory experiments, with **raspberry grey mould,** *Botrytis cinerea,* have shown that relative humidity changes, both increases and decreases, lead to jerky movements of the spore-bearing stalks, and can account for the rates of shedding that were studied for five years in the field in Scotland (Jarvis 1962, Fig. 4). Spore shedding in both rising and falling relative humidities has also been studied with the fungi causing **rice blast,** *Pyricularia oryzae* (Leach 1980a) and **northern leaf blight of maize,** *Drechslera turcica* (= *Helminthosporium turcicum*), by Meredith (1965) and by Leach (1976, 1980 b and c). There was evidence to support an electrostatic mechanism for violent, active discharge. It is supposed that evaporation

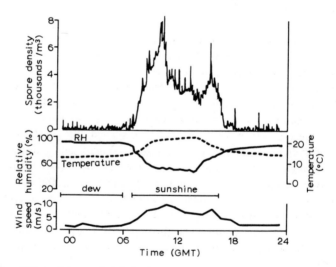

Fig. 4   Effect of changes in relative humidity and temperature on release rate of spores of raspberry grey mould, *Botrytis cinerea,* from a crop in Scotland, 3 August 1960. The dew dried rapidly in the sun and wind. (After Jarvis 1962).

accompanying falling humidity leads to a build-up of charge, and a force that repels like-charged spore and support grows until it is greater than the force joining spore to its support, when the spore is shot off. Such a mechanism might well be more general. Other fungi have a kind of sling in which the drying, straining cell next to the spore suddenly changes shape as a bubble forms within it.

Mosses grow their spores in capsules. In most species the capsules simply open with falling relative humidity, but in the **bog-moss,** *Sphagnum*, an air space below the spore mass is squeezed until the capsule suddenly breaks open and the spores are shot out.

Most **ferns** throw off their spores from minute capsules that split in dry air. As the wall cells of a split capsule dry out further, they become greatly strained until there is a sudden return to their former shape (caused by bubble formation in the cell water), and spores are flung out as the capsule snaps shut. **Leafy liverworts** throw off their spores in much the same way — from the outside of a capsule that shrinks in dry air and then suddenly regains its shape as a bubble forms inside.

1.1.3 Wind
Active take-off by some **wingless insects** is encouraged by light winds. First-stage caterpillars of the **gypsy moth,** *Lymantria dispar,* a pest of many species of woodland and forest trees, have a mass of less than 1 mg at hatching. At night, or in cool or rainy weather, they stay on the underside of leaves; otherwise they move about and feed, but when crowded some arch their bodies and let go of the leaf. Their rate of fall is lessened by a covering of long hairs and by silk threads that they let out and below which they hang (page 92). Field studies in Connecticut oakwoods showed that the first caterpillars became airborne on the day after hatching started, and that greatest catches were in the morning, after the threshold temperature had been passed (Leonard 1971). Similar field studies with the **Douglas-fir tussock moth,** *Orgyia pseudotsugata,* in the forests of British Columbia, showed that it, too, took off almost wholly as first-stage caterpillars soon after hatching, and again the peak rate was during late morning on warm, dry, sunny days with winds no stronger than about 2 m/s (Mitchell 1979). Second-stage caterpillars of the **eastern spruce budworm moth,** *Choristoneura fumiferana,* also drift on silk threads after over-wintering, dropping on their threads and swinging until they break in a strong enough gust (Batzer 1968).

Some **spiders** are also well known to be carried by the wind on silken threads of gossamer. In laboratory studies of **wolf spiders** of the genus *Pardosa*, 5- to 10-day-old first-stage larvae, 1.5–2.0 mm long, were warmed to 30–35°C and ventilated by a fan (Richter 1967). Some of them spun threads up to 70 cm long before releasing their hold, but this took place only in wind speeds 0.8–4.0 m/s. In later studies with *P. purbeckensis* (Richter 1970), the youngest stages were the most active at 'tip-toeing' and taking off on threads, and tip-toeing increased with rising temperature and falling relative humidity. Hence take-off is most likely to happen on warm, dry days with light but gusty winds. Vugts & Wingerden (1976) found that the fraction of *Erigone arctica* tip-toeing on the tops of grass decreased with increased wind speed (Fig. 5).

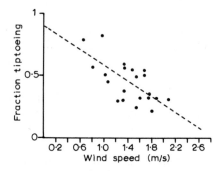

Fig. 5 — Effect of wind speed variation (at height of 20 cm) on the fraction of a population of the spider *Erigone arctica* tip-toeing at the tops of a grass stand. (After Vugts & Wingerden 1976).

Like spiders, some species of tetranychid **mites** lower themselves from host plants on silk threads that break in the wind. For example, larvae of the **European red mite,** *Panonychus ulmi,* a pest of decisuous fruit trees, use threads up to 10 cm long, and the same is true for the **citrus red mite,** *P. citri,* a pest of citrus and some other trees (Jeppson *et al.* 1975). By contrast, adults of the eriophyid, *Eriophyes tulipae,* the **wheat curl mite,** a carrier of **wheat streak mosaic virus,** swarm up stems when they become weakened by heavy infestation. At the top they form seething masses, often in chains that break away in gusts (Gibson & Painter 1957, Nault & Styer 1969). Laboratory experiments with the **mite** *Amblyseius fallacis,* which feeds on other mites that are pests in apple orchards, have shown that it is the females that are most prone to becoming airborne, just before or during egg laying (Johnson & Croft 1976). They face into winds stronger than about 0.5 m/s, until only the rear pair of legs is clinging, and then they let go their hold. By contrast, Lanier & Burns (1978) have suggested that **bark beetles** (Scolytidae) actively *avoid* take-off in gusty winds by reacting to the accompanying fleeting fluctuation of **atmospheric pressure,** sensed by a corresponding shrinking and swelling of air bubbles they had swallowed. It is supposed that gusty winds make it difficult for the beetles to follow windborne scent trails (page 101).

Turning to active take-off by **winged insects,** laboratory studies with groups of about 40 **blowflies,** *Calliphora erythrocephala,* standing in a small wind tunnel have shown how take-off rate varies when wind speed suddenly changes from 0.5 m/s (Digby 1958). Changes to both stronger and weaker winds led suddenly to lower take-off rates, whereas change back suddenly to 0.5 m/s have higher take-off rates. Changes to or from a stronger wind were followed by a slow return to the rate for 0.5 m/s winds (over 10–12 minutes – greater for longer spells at the strong wind), whereas changes to or from a weaker wind were followed by almost no trend to go back to the rate for 0.5 m/s (Fig. 6).

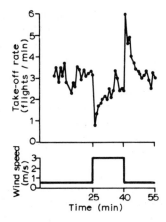

Fig. 6 – Effect of sudden wind speed changes, from 0.5 to 3.0 m/s and back again, on take-off rate in a group of 40 blowflies, *Calliphora erythrocephala*. (After Digby 1959).

Take-off was most rapid for 0.7 m/s winds. These studies hint at two conflicting ways in which flies behave in a suddenly stronger wind: the first leads to a greater take-off rate whenever the wind strengthens, no matter what its speed is; the second cuts down take-off rate when a strong wind suddenly strengthens even more. Lulls therefore lead to an increased take-off rate during fitful winds in the field. There seems little doubt that the **blowfly** *Calliphora vicina* is encouraged to take off by falling **atmospheric pressure**. Edwards (1961) used the fact that when a fly takes off it picks up a small electrostatic charge. He studied take-off rate by means of an electrometer in a cage wired to earth. After keeping the flies at constant pressure for a day, the pressure was made to fall steadily. Take-off rate grew as pressure fell, whereas it had not grown when the pressure was steady. Sudden pressure drops (1 mbar in 15 min) caused noteworthy jumps in take-off rate.

Another example of the effect of wind on active take-off by **winged insects** is provided by the **red locust**, *Nomadacris septemfasciata*, often scattered across those grassy plains of East Africa that are liable to seasonal flooding. In a study of take-off during fitful winds, Chapman (1959) noted the numbers of locusts flying across fixed lines in the Rukwa Valley of Tanzania. Take-off was found to be thwarted by strong winds but helped by light winds. Numbers flying were greatest about 2 min after the first dropping of the wind below 1.5 m/s, a time lag that hints of take-off being helped by a temperature rise resulting from reduced cooling during light winds. The inside body temperature of a locust in sunshine, pinned above a wind vane such that it always lay broadside to the wind, rose by about 1°C in the first 2 min of light wind, but a greater rise can be expected in the shelter of vegetation. Similarly, the **Australian plague locust,** *Chortoicetes terminifera,* was also found to take off during lulls in strong, gusty winds (Casimir & Bament 1974), and so did the **European elm bark beetle,** *Scolytus multistriatus* (Meyer & Norris 1973). The apparent inconsistency of the latter with the findings of Lanier & Burns (page 20) may be due to observations having been made at different stages in the beetle's life cycle.

Insects usually take off into wind. This is aerodynamically sensible, and has been demonstrated for a wide range of species, including **mosquitoes** (Klassen & Hocking 1964), **aphids** (Haine 1955), **weevils** (Solbreck 1980), **thrips** (Lewis 1973), **locusts** (Roffey 1963) and **lacewings** (Duelli 1980b). Once airborne, forward movement may be lost straightaway or after climbing to above the insect's boundary layer (page 95).

Many kinds of organisms are *passively* knocked or shaken into the air by wind gusts. For example, the **spores** of some **fungi** (rusts, smuts and powdery mildews, among others), **lichens, mosses, liverworts** and **ferns,** and the **pollen, seeds** or **fruits** of many kinds of flowering plants, are carried on stalks well exposed to wind gusts. The fungus *Helminthosporium maydis* causes the disease known as **southern corn leaf blight.** Specimens of race T were grown on 60-day-old corn, and pieces of infected leaf were placed in a wind tunnel (Aylor 1975a). Using a small air jet it was found that wind speeds of 8–12 m/s led to the most rapid removal of spores (Fig. 7), equivalent to a drag force of 1×10^{-2} dynes on a single spore. Spores tended to come off together, suggesting that those released struck others downwind still fixed. On the leading edge of a leaf, as many as 90% were released after 3–5 min, even with wind speeds only 6–7 m/s — presumably because the viscous boundary layer (page 48) was not fully developed at the leading edge, and spores there poked through it. Such strong winds,

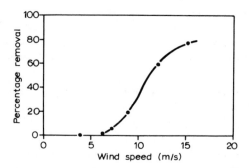

Fig 7   Effect of wind speed on percentage removal of spores of southern corn leaf blight *Helminthosporium maydis,* reared on dry plants. Means of 4 to 6 measurements at each wind speed over 15 sec periods. (After Aylor 1975a).

however, are unlikely to be the main cause of spore take-off. Indeed, field observations have shown that spores are shed in 1 m/s winds at crop top, particularly after the leaves have dried out (Aylor & Lukens 1974). Fluttering of leaves, rather than their rubbing together, is the most likely mechanism (Aylor 1976, 1978). Experiments with 2-5-week-old seedling wheat plants in a small wind tunnel showed that the releasing agent for spores of the **cereal mildew** *Erysiphe graminis* was gusts (Hammet & Manners 1974). Further laboratory work with infected barley leaves shaken mechanically showed that spores could be shaken off by accelerations of 0.4 m/s² (Fig. 8). Speeded-up cine photography of barley plants 50 cm tall and with three large leaves showed that when they

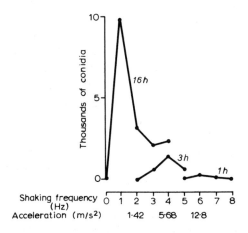

Fig. 8 – Effect of frequency of shaking of barley leaves on release rate of spores of cereal mildew, *Erysiphe graminis,* after rest periods of 1, 3 and 16 h. (After Bainbridge & Legg, 1976).

were shaken in a wind of only 0.6 m/s there were accelerations greater than 1 m/s² (Bainbridge & Legg 1976). It seems that shake-off is a main release mechanism for spores of fungi growing on exposed parts of plants. Sutton *et al.* (1976), from field experiments near an orchard in Michigan, found that spores of *Spilocea pomi* (the imperfect stage of the fungus causing **apple scab)** were spread in warm, dry, sunny and windy weather, probably after taking off by the rubbing together of infected leaves. It is possible that the bacterium, *Erwinia amylovora,* the cause of **fireblight** in apples and pears, can become airborne in wind gusts. On young, rapidly growing, infected shoots, bacteria can appear in 25 μm diameter strands up to 1 cm long that exude from the tissues (Bauske 1968). When dry they curl up like springs.

By far the greatest bearers of **pollen** are **conifers.** In early summer, clouds like yellow smoke can at times be seen drifting from forests, the floors of which have an easily seen carpet. Among **flowering plants,** however, there are many more species whose pollen is carried by insects than by the wind. Grasses form the largest group of wind-pollinated plants. Grass pollen is the commonest cause of **hay fever** and **skin allergies.** As little as one pollen grain in 1 m³ of air can set off hay fever. Rushes (Juncaceae) and sedges (Cyperaceae) all give off much windborne pollen, but it does not cause hay fever. Windborne pollen is also given off by many other kinds of flowering plants, but most shed too little for it to cause much disease. Amongst wind-pollinated herbs, mostly summer and autumn flowering, are some Compositae (notably ragweed *Ambrosia,* false ragweed *Franseria,* and mugwort *Artemisia*) and some Chenopodiaceae (notably beet *Beta vulgaris,* but also Russian thistle *Salsola,* burning bush *Kochia,* and saltbush *Atriplex*), as well as *Plantago, Rumex* and *Urtica.* Among wind-pollinated trees and shrubs, often spring flowering, are *Betula, Alnus, Corylus, Platanus, Fagus, Quercus, Ulmus, Morus, Broussonatia, Juglans, Corya, Populus, Acer, Fraxinus* and *Ligustrum.* Flowers of wind-pollinated plants tend to be small and placed at the ends of branches, where they often open before the leaves (as in the catkins of many middle-latitude trees). The two sexes may be on different plants, or if they are on the same plant they are often in different flowers that open at different times, thus avoiding self-pollination. Anthers stand out on slender stalks or in catkins, and their jostling in gusty winds causes pollen shedding, most often on warm, dry afternoons. Stigmas are large and well exposed, and are often divided or feathery so as to increase the chance of catching windborne grains. Small or no petals aid the take-off and landing of windborne pollen. But if one effective grain is to reach a 1 mm² stigma, every 1 m² of plant habitat needs something of the order of 10⁶ pollen grains. In fact, production is ample to reach this kind of density: each catkin of *Betula* or *Corylus,* for example, produces several million grains (Proctor & Yeo 1973).

Similar to the jostling anthers shedding their pollen are the **fruits** shaped like pepper shakers on long stalks, with holes through which seeds are shed when a fruit is knocked by a wind gust. Most mosses have spore capsules of this kind. The poppy, *Papaver,* is a well-known example amongst flowering plants. Sometimes a part or the whole of a dead plant breaks away and rolls along the ground in the wind as a tumbleweed, shedding its seeds as it goes. Many species from windswept open places behave in this way.

Some **seeds** are so small and dust-like (for example, those of Orchidaceae) that they behave like spores or pollen, but most are heavy enough to go no further than a few tens of metres through the air. There are some, however, that have devices for lowering their rate of fall. Three kinds of such devices can be recognised: balloons, plumes and wings. Somewhat like the loose bundle of a tumbleweed, the balloon-like seed pods of *Colutea* break off and blow away on the wind. Other seeds are hairy (notably cotton, *Gossypium*), and so are some small fruits. Other fruits have a feathery outgrowth (pappus) so that they behave like parachutes once airborne (notably some Compositae, like the dandelion, *Taraxacum*).

Some flowering plants actively throw their seeds into the air in ways like those used by fungi. A strain can build up in the seed coat, or in the fruit as a whole, and it may be released by a gust of wind.

Loose, dry **soil**, infected with minute organisms, can be blown into the air by the wind. The mechanisms are still unclear, but it seems that the larger particles roll or bounce along, strike smaller particles in their path and thereby knock them into the air. **Nematodes** become windborne in this way (Orr & Newton 1971). Some farming methods also release spore-laden dusts that on being breathed in deeply lead to illnesses known as 'farmer's lung'.

1.1.4 Rain

An example of triggering the throwing of **seeds** into the air by rain is provided by *Blepharis ciliaris*, a plant able to live in the dry parts of north-east Africa and south-west Asia. After wetting, a seed capsule explodes violently due to release of a strain set up by drying out of some parts of the capsule (Gutterman *et al.* 1967). Some plants shed their seeds on the first of the fitful rains characteristic of its habitat, but others do so only after a long soaking; hence shedding is spread through the rainy season. Lying rain, or even dew, can bathe spores, pollen and even small insects in water, which must evaporate before they can take off. Moreover, the evaporation can cool insects enough to hinder take-off. On the other hand, rain can trigger the flight of larger insects within hours (for example, termites) or a few days (for example, *Phyllophaga crinita,* a **scarab beetle** pest of grass and crops, whose egg survival is poor in very wet soil (Gaylor & Frankie 1979)).

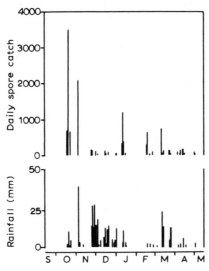

Fig. 9 – Daily rainfall and counts of ascospore octads of the apricot pathogen *Eutypa armeniacae* caught by sampling 0.6 m^3 air/h, 1.8 m from a spore source on a dead branch in the Suisun Valley, California, 1970–71. (After Ramos *et al.* 1975).

Some fungi release their spores only after wetting, presumably when the air around the fruiting bodies becomes saturated (page 35). *Eutypa armeniacae* is a pathogen of apricot, *Prunus armeniaca.* Disease is widespread in drier, irrigated areas of California. Spores are released in large numbers during the first autumn rains, but only if the fall is greater than about 1 mm (Ramos *et al.* 1975, Fig. 9). Field studies with species of **tremellaceous fungi** showed that spore shedding started within two hours of wetting by rain, and continued until they dried out — which was delayed by the presence of a substrate which was large compared with the fruiting body, and therefore when wet able to keep a high air humidity (Rockett & Kramer 1974). Ascospores of *Eutypella parasitica,* the cause of **stem cankers in maple** (*Acer*), were found from measurements in woodlands in Wisconsin after rain to be ejected up to 9 mm (Johnson & Kuntz 1979, Fig. 10).

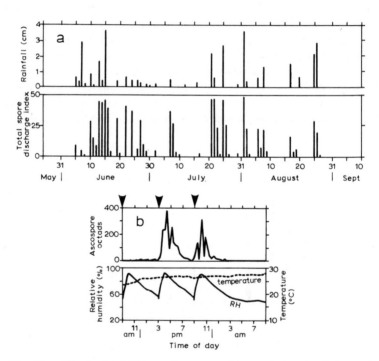

Fig. 10 — Effect of rainfall on release rate of ascospore octads of the maple stem canker fungus *Eutypella parasitica.* (After Johnson & Kuntz 1979).

  a Daily rainfall and discharge from 10 cankers in a Wisconsin woodland during 1967, expressed as the sum of *discharge indexes* in 17 spore traps placed a few millimetres away. Discharge index is defined subjectively as — 0 = no spores, 1 = few, 2 = single layer, 3 = two or more layers.

  b Changes in release rate induced by water sprayed at 6 h intervals at the times shown by arrow heads. Air was sampled 1 m from a canker in a closed chamber.

A very effective way of getting some kinds of fungal **spores** into the air is being knocked by individual raindrops (or drip drops within a plant canopy). Four mechanisms have been discovered.

Blowing — a bellows-like fruiting body is suddenly pushed out of shape by the falling drop, and spore-laden air is blown out.

Tapping — a leaf is made to flutter by a drop and it sheds spores from the still dry parts.

Puffing — a very short-lived gust spreads outwards from the striking point, strong enough to carry away spores from the still dry parts.

Splashing — droplets spraying outward from the striking point carry away spores or patches of a spore-laden film of water.

Gregory (1949) let 5 mm diameter drops fall on to the large, spore-filled fruiting bodies of the **puff ball fungus,** *Lycoperdon perlatum,* and the resulting spore cloud blown through the hole at the top was filmed by very high speed cine photography. Spores were found to leave 3 ms after a drop struck, but the cloud had almost come to a stop after another 30 ms, having travelled about 2 cm. The smallest drops needed to form a cloud were about 1 mm diameter, a size that is common in most rainfalls (page 42). That both tapping and puffing are involved in the knocking into the air of spores of **wheat rust,** *Puccinia graminis,* was shown in experiments by Hirst & Stedman (1963). They let both water drops and glass beads fall on rusted wheat straws, and found that drops set free more spores than did the beads. Although both drops and beads set spores free by tapping, drops seemed to be using another mechanism as well. This suspicion was confirmed by using straws fixed to bendable plastic strips, for fewer spores took off when a straw was on the underside of a strip, where no puffing is possible. Using spores of the **club-moss** *Lycopodium* on an unbendable iron block, they showed that few took off when they were struck by beads, but many spores took off when they were struck by drops and some of them were left as rings, as would be expected from sudden outward-spreading puffs of air. The splashing mechanism was examined by Gregory *et al.* (1959) using a wetted twig of sycamore, *Acer pseudoplatanus,* infected with the **coral spot fungus,** *Nectria cinnabarina*. The twig was held 10 cm above an array of sampling slides placed in rows and columns out to 50 cm. Drops of 5 mm diameter falling on the twig at about terminal speed (page 57) each gave about 2,600 splash droplets, which fell on to the slides and left clear spots in the coating of dyed gelatin (Fig. 11). Less than 1 droplet/cm^2 fell beyond 15 cm. All the droplets carried spores. Similar results were obtained by Dowding (1969) using 3 mm drops falling 3 m on to cultures of *Ceratocystis* species (causing **blue-stain** in pine trees); large numbers of spores were caught up to 2 m away. Rain splash droplets caught downwind of coffee bushes infected with the fungus causing **coffee berry disease,** *Colletotrichum coffeanum,* were found to contain spores of the fungus, but many fewer than in the water running down branches and dripping from berries (Waller 1972).

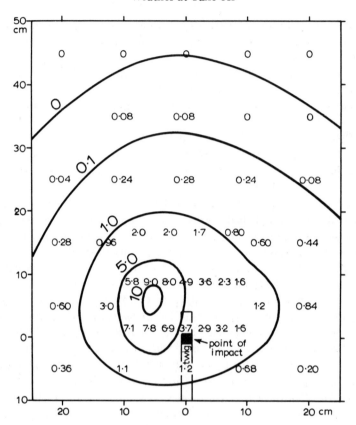

Fig. 11 – Pattern of splash droplets around the point of impact of a 5 mm diameter drop falling at about terminal speed from a height of 7.4 m on to a wetted
sycamore twig infected with coral spot fungus, *Nectria cinnabarina*. Numbers of
droplets/cm² of splashed area. (After Gregory *et al.* 1959).

The effects of splash from single large drops have been examined in the
laboratory and over mown grass (Stedman 1979), as well as in growing field
beans and wheat (Stedman 1980a and b), with the aid of fluorescein in the
water. In the laboratory, splash droplets from 3 to 4 mm diameter drops falling
on to a 0.1 mm deep water film were found to contain about equal parts from
drop and film, and there was a linear relation between the volume splashed and
kinetic energy of the drop. In the field, dye was detected only as far as a few
metres (Fig. 12), but would have been taken much further in rain, because of
repeated splashing. With wheat, splash droplets spread further when the source
within the crop was higher. Rain in winter, with stronger average winds and a
shorter and more open crop, would lead to a rapid spread of spores; whereas in
spring and summer, with lighter average winds and taller crop, many drops
would be broken by the upper canopy before being able to reach and strike

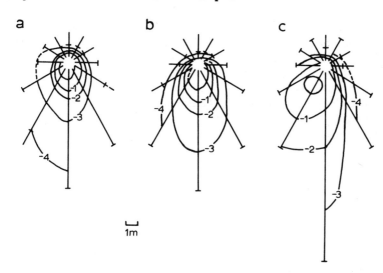

Fig. 12 – Examples of splash patterns at ground level around the point of impact of 200 ml of fluorescein solution in 4.4 mm diameter drops falling at about terminal speed from a height of 7.6 m on to a free-draining target at three heights above mown grass – a 0 cm, b 50 cm, c 100 cm. Isopleths on a scale of $\log_{10}$ (μl/cm^2). Radii show distances to which measurements were made. Ticks on the radii show distances to which fluorescein was detected. The wind presumably blew in the direction from top to bottom. (After Stedman 1979).

diseased parts, which are mostly in the middle and lower canopy. Moreover, light winds within the denser crop would lead to little spread. Nevertheless, heavy summer rains in light winds would have large drops falling vertically and more easily able to reach diseased parts.

Similar experiments with drops falling on to lengths of senescent potato stems collected from a potato crop with **gangrene,** caused by *Phoma exigua,* var. *foveata,* showed that aerial density of spores increased with time to a maximum, depending on drop size and potato cultivar (Carnegie 1980). Some spores were being released up to an hour after the start. The first appearance of airborne spores in association with rain has been demonstrated in the field for several fungal disease organisms, including *Rhynchosporium secalis,* the cause of **barley leaf blotch** (Stedman 1980c); *Mycosphaerella graminicola,* the cause of **speckled leaf blotch of wheat** (Brown *et al.* 1978); *Puccinia striiformis,* the cause of **yellow rust** on both barley and wheat (Rapilly 1979); *Eutypa aremeniacae,* a pathogen of apricot (Ramos *et al.* 1975). and *Botrytis squamosa,* the cause of **leaf blight on onion** (Sutton *et al.* 1978). Rate of spore release will vary not only with source size but also with intensity and duration of rain as well as drop size. Much the same will be true of overhead irrigation water. The effectiveness of rain splash in spreading barley leaf blotch has been demonstrated by means of field sampling: numbers of spores caught increased with rainfall intensity (Stedman 1980c). Even drifting fog droplets, despite their small size (often less than 10 μm diameter) can knock some small spores from their stalks by means of the so-called 'mist pick-up'. Laboratory experiments with *Ceratocystis* have demonstrated this (Dowding 1969). Spores were found to be easily dislodged when colonies of the fungus were put in a wind tunnel and subjected to a wind charged with droplets 2-60 μm diameter (produced from the edge of a spinning disk). Rate of take-off increased more or less linearly with wind speed from 0.5 to 10 m/s.

Rain splash is known or thought to be the mechanism of making windborne the spores of fungal diseases in a great variety of crops including:

Fungus	Cause of
Phytophthora infestans	**late blight of potato**
Septoria lycopersici	**leaf spot of potato**
Mycosphaerella berkeleyii	**leaf spot of peanut**
Erysiphe polygoni	**powdery mildew of bean**
Isariopsis griseola	**angular leaf spot of bean**
Mycosphaerella musicola	**leaf spot of banana**
Colletotrichium musae	**anthracnose of banana**
Cercospora coffeicola	**brown eye spot of coffee**
Ceratocystis fimbriata	**canker of coffee**
Exobasidium vexans	**blister blight of tea**
Monilia roreri	**watery disease of cocoa**

In some plants, spores are produced inside vase-shaped structures, pointing upwards and with circular mouths of diameter 5-8 mm. Brodie (1951) has confirmed that when a raindrop falls inside, the spores are shot outwards a metre or more in the splash droplets (Fig. 13). These **splash cups** have opening diameters a little greater than the largest raindrops (page 42). The **birds-nest fungus,** *Cyathus striatus,* and the **liverwort** *Marchantia polymorpha* are examples of plants whose spores are shed in this way. Soredia of the **lichen** *Cladonia* are also thrown from splash cups.

Seeds of some flowering plants are also thrown into the air by falling

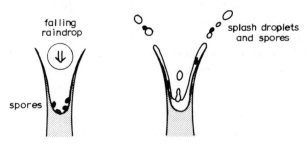

Fig. 13 – Schematic splash cup, from which spores are forcibly thrown by a splashing raindrop. (After Brodie 1951).

raindrops. Some plants have more or less horizontal seed pods which, when open, have their seeds knocked out. The plant *Salvia lyrata* has a spring-board mechanism. Its fruits are nutlets at the base of a more or less horizontal tube-like calyx that is fixed to the main stem by a springy stalk. A raindrop hitting its upper side bends the tube down. On springing back, the fruits are shot out up to 2 m, guided by a groove in the lip of the lower part of the tube (Brodie 1955). The plantlets (gemmae) along the leaf edges of *Kalanchoë (Bryophyllum) tubiflora* can be thrown similarly up to 2 m (Brodie 1957).

Viruses and **bacteria** can also be forcibly launched by splashing. Drops falling into infected water give bubbles that, on bursting, throw off small droplets as the tiny pit originally formed in the water surface falls inward and gives a jet that rises from the middle and breaks. Bursting bubbles are also formed in running water, and when air is bubbled artificially through fluids, for example, sewage in treatment tanks. In laboratory experiments with bursting bubbles in a water suspension of the conveniently red, and therefore easily seen, rod-shaped bacterium *Serratia marcescens,* about 1 μm long, droplets were collected on culture plates to show the density of bacteria in each (Blanchard & Syzdek 1970). The density was found to vary with droplet size (radii 10-70 μm), being greatest at 30-40 μm, where it was about 1000 times the density in the suspension as a whole, most likely because the bacteria had been gathered into the skin of the droplets. The bacterium *Erwinia amylovora* seems to spread in splash droplets, as well as possibly by strands (page 23). Simulated wind and rain in greenhouse experiments caused it to become airborne (Bauske 1967). Related species of *Erwinia* cause **soft rots.** Laboratory experiments with 3 and 5 mm diameter drops falling on pieces of potato stem infected with *E. carotovora* var. *atroseptica,* the cause of **blackleg,** showed that bacteria were taken away in the splash droplets, but only a very small fraction of those available (Graham & Harrison 1975, Graham *et al.* 1977; see also Pérombelon & Kelman 1980).

Infected droplets can also be formed by sudden break-up of films of watery mucus when animals breathe, cough, sneeze or defecate, especially when the animals have diseases of breathing tubes or gut. **Foot-and-mouth** is thought to get into the air this way by breathing, especially from pigs (Sellers & Parker 1969). Man makes his own contribution to these mechanisms that shed infected droplets into the air: for example, by sound vibrations when talking, by the hosing down of floors, by the flushing of water closets, by the spreading of farm slurry on fields, and by the overhead watering of crops. In Israel, comparisons of the incidence of **enteric communicable disease** in 77 settlements that used wastewater irrigation (not for food crops) with 130 that did not use it, showed a two to four times increase of the disease in the former during summer but not during winter (when there was no irrigation). This suggests windborne spread of disease (Katzenelson *et al.* 1976). Air samples taken downwind from sprinklers using treated sewage, and marked with tracer bacteria, demonstrated that spread did occur (Teltsch & Katzenelson 1978).

1.2 WEATHER ELEMENTS

The previous section has given examples of how weather changes affect the take-off of organisms. To get some understanding of these changes, it is helpful to consider next the nature of the principal weather elements: temperature, humidity, wind and rain. We can then proceed (in the remaining sections of this Chapter) to a discussion of the causes of weather changes. That discussion will also help when considering the effects of weather on landing (Chapter 2) and windborne spread (Chapters 3-7). The great variety of weather changes, and their causes, are considered extensively in many books on meteorology, such as Geiger (1965), Monteith (1973 and 1975), Yoshino (1975), Grace (1977), Oke (1978), Scorer (1978), Pedgley (1979) and Ludlam (1980). Here we look only at those principles that are of most direct concern to an understanding of atmospheric effects on take-off by airborne organisms.

1.2.1 Temperature

We need to consider the temperature of the air, of the organism, and of the surface upon which the organism may be settled or growing (such as the ground or vegetation). *Air* temperature can be measured by a thermometer exposed to the air in such a way that it is shielded from such heat sources as sunshine, or radiation from warm ground. The temperature of an *organism* or *surface* can be measured either directly, by a thermometer exposed in such a way that it has the same temperature as that of the organism or surface; or indirectly, with a thermometer that measures the intensity of infra-red radiation coming from the organism or surface. The temperature of a very small organism can be difficult to measure, but it may be taken as the same as that of the air around it, or that of the ground or vegetation on which it lies.

In sunshine, an *airborne* organism will be warmer than the air, whereas on a cloudless night it will be a little cooler. The temperature difference is less for smaller organisms, and for most will be less than 1°C. Insects in flapping flight, however, generate heat in their wing muscles, so they are always warmer than the air, even at night, and for large species the difference can be more than 5°C.

In sunny weather, the ground or vegetation, and the small organisms to be found on them, may be several °C warmer than the air that is only a few centimetres away, whereas on clear, quiet nights the reverse can be true. These differences tend to be less in windy weather, when conductive heat flow to the air is greater. Temperature can also vary markedly over a surface, according to its inclination to the sun's rays and exposure to the wind — for example, over the surface of a tree trunk, or that of a growing apple, or of a stone lying on the ground, or even from one side of a valley to another. Differences as great as several °C can occur even across a single leaf. Cloudy skies reduce these differences because clouds cut down the intensity of both incoming short-wave (visible) radiation from the sun and outgoing long-wave (invisible) radiation from the ground, the vegetation and the air. Such differences should be borne in mind when considering the effects of temperature on organisms taking off in fitful sunshine, and in cases where only the *air* temperatures can be measured. In dull, windy weather there are only small spatial differences in temperature of ground, vegetation and air.

The atmosphere is largely transparent to sunshine; it is heated and cooled by contact with the ground, and by exchange of long-wave radiation between it, the ground and outer space. When the ground is warmer than the air, convective mixing carries heat upwards through a **convective layer** that is very variable in depth — sometimes as much as several kilometres. Thorough mixing brings about a decrease of temperature upwards (**lapse rate**) at about 10°C/km (equal to the dry adiabatic lapse rate, DALR). This decrease is caused by: (a) cooling inside the convective updraughts as they expand on rising into lower pressure aloft, and (b) mixing of the updraughts with their surroundings. The atmospheric lapse rate is greater close to the ground, where radiation and conduction exceed convection as the means for upward exchange of heat. For heights greater than a few metres the DALR can be used to get a good estimate of air temperature at any height within the convective layer. For example, at a time when the layer is known to be 2 km deep (from information available from national meteorological services) and the near-surface air temperature is 25°C, then at 500 m above the surface it will be about 20°C, at 1 km about 15°C, and at 2 km about 5°C. Such large temperature differences can affect the height of insect flight (page 191). Above the convective layer, the lapse rate is less than adiabatic, and moreover it will vary with place and time. Hence atmospheric temperatures aloft are best interpolated from soundings made by instruments that are either carried aloft by balloon, aircraft or satellite, or placed on tall towers.

On a cloudless night, especially when the wind is light, the ground and vegetation tend to be cooler than the air, so heat in the lower atmosphere flows downward. Air temperature then *decreases downward,* and there is a **temperature inversion,** even as much as 10°C in 100 m. By dawn, the warmest air is usually at a height of several hundred metres (Fig. 14a). A difference of 10°C can develop between top and bottom of an inversion on a clear, still night — a difference that can greatly affect the height of night-flying insects (page 191). Night-time temperature changes above the inversion are much smaller than changes near the ground, so a first estimate of night-time temperatures there would be the same as those for the previous afternoon, based on the maximum temperature measured at about 1 m and assuming an adiabatic lapse rate. For example, if the maximum was 30°C and even if the 1 m temperature had fallen to 15°C by 3 a.m. in the morning, air temperature at the top of the night-time inversion (taken to be at a height of 500 m) will be about 25°C. Such night-time inversions are weak or absent on cloudy or windy nights. Estimating temperature changes and differences in this way is unreliable, however, when there is a wind that can bring in warmer or cooler air from elsewhere.

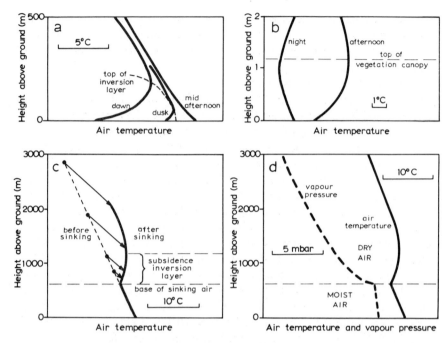

Fig. 14 – Typical air temperature profiles (lapse rates).
  a With clear sky and light wind, showing deepening of a temperature inver-
      sion during the night from the ground upwards (the difference between
      the dawn and mid afternoon profiles shows the variation of diurnal range
      with height);
  b Within a vegetation canopy.
  c With warming due to sinking, showing formation of a temperature inver-
      sion with its base above the ground (arrows show temperature changes due
      to sinking from four sample heights).
  d Same as c, but showing dryness of the subsided air in contrast to the moist-
      ness of the air below.

Inside a vegetation canopy, highest day-time and lowest night-time temperatures are within the upper part of the canopy. Hence, there tends to be a temperature inversion in the lower canopy by day and in the upper canopy by night
(Fig. 14b).

    Temperature changes are caused not only by changes in the *local* heat
balance but also by the arrival of warmer or cooler air from elsewhere. Such
advection by the wind can be horizontal or vertical, and it depends upon there
being a wind blowing in the presence of a **temperature gradient**. Changes can be
rapid where there are strong winds and a large gradient (page 46). *Horizontal*
temperature gradients arise from differential heating and cooling of the earth's
surface. This can happen on a wide range of scales – for example, between poles
and equator, between land and sea, between bare soil and vegetation, between
air chilled by evaporation of falling rain and unchilled air. *Vertical* gradients

(lapse rates) of less than 10°C/km arise where there is poor vertical mixing, as often happens above the daytime convective layer. Air sinking in the presence of such a lapse rate brings about a warming at any given height, whereas air rising brings about a cooling. A greater sinking rate at greater heights can lead to a temperature inversion (**subsidence inversion**) aloft (Fig. 14c), unlike the night-time inversion that starts at the ground. Because both wind and horizontal temperature gradient can vary with height, temperature inversions can also form aloft when warm air flows over cold air (**frontal inversion**). At any one time, there can be several inversions above a given place, and any one of these inversions can have one or more origins.

1.2.2 Humidity

Air always contains some water in the form of invisible **vapour.** Although the proportion of water vapour relative to the other air constituents is very variable, it seldom exceeds 3%. A measure of the water vapour content is the air **humidity,** which can be expressed in several ways, notably:

> **vapour pressure** — that part of the total atmospheric pressure due to the water vapour.

At sea level, total pressure is about 1000 mbar, and the water vapour can contribute as much as 30 mbar. Air is said to be **saturated** when its water vapour content cannot be increased by evaporation at constant temperature. The proportion of vapour in saturated air and the **saturation vapour pressure** both increase with temperature. Hence, the cooling of saturated air leads to condensation, whereas the warming of saturated air leads to unsaturation and the possibility of evaporation. Sufficient cooling of unsaturated air will lead to saturation, and the temperature at which this happens (at constant pressure) is the **dew point.** The degree of unsaturation can be expressed in several ways, notably:

> **relative humidity** — the *ratio* of actual and saturation water vapour pressures at the given air temperature;
> **saturation deficit** — the *difference* between actual and saturation water vapour pressures at the given air temperature;
> **dew point depression** — the difference between dew point and air temperature.

Whereas both water vapour pressure and dew point of a given mass of air increase (decrease) only by adding (taking away) some water vapour, the *relative* humidity can change not only in this way but also by cooling (heating) the air.

Water vapour gets into the air by **evaporation,** mostly from liquid water in the ocean, soil and vegetation; and it leaves the air by **condensation,** mostly through clouds and rain. Most rain falls on the oceans or reaches them through rivers; the rest is taken up by soil and vegetation. Water vapour is mixed into the lower atmoshere by wind gusts (page 47) and then taken upwards to greater heights in deeper and often cloudy updraughts, especially within large convective clouds and near atmospheric fronts (page 52). The combined effects of evaporation at the earth's surface, and condensation in clouds causes an upward decrease of dew point, known as a **hydrolapse.** In well stirred but unsaturated air, the hydrolapse is small; by contrast, in poorly stirred air, water vapour may spread only very slightly upward or downward, and the hyrolapse is then large. For example, a temperature inversion can trap water vapour beneath it, and the dew point *decreases* rapidly upward through the inversion. When the ground is colder than the dew point of the air, and water vapour is being condensed out of the air in the form of dew or fog, the dew point *increases* upwards near the ground. Within a plant canopy, the hydrolapse is more complex because there can be evaporation from, and condensation on, both soil and vegetation.

Humidity changes are due not only to *local* evaporation and condensation but also to **advection** of moister or drier air from elsewhere — by wind in the presence of a humidity gradient. Horizontal gradients of humidity are brought about by differential evaporation over the earth's surface, and also by the dependence on temperature of saturation vapour pressure. As with temperature, horizontal gradients occur on a wide range of scales. Vertical gradients arise when mixing is not thorough, and are common. Water vapour pressure (and dew point) can either increase or decrease upward: the variation of hydrolapse is strongly affected by differential horizontal advection of vapour (due to variation of wind and horizontal humidity gradient with height) and to differential vertical advection. For example, the air above a subsidence inversion often has a small water vapour pressure because it has sunk from heights where temperatures were so low that even saturation vapour pressures were small (Fig. 14d). *Relative* humidity changes are more complex because they depend upon changes in both temperature and vapour pressure. For example, in thoroughly mixed air, where the vapour pressure is constant with height, the relative humidity increases upward, because temperature decreases upward.

1.2.3 Wind
Wind is air moving over the ground; it has both speed and direction. Although the wind is in general three-dimensional, its vertical component is about two orders of magnitude smaller than the horizontal component (except on certain occasions mentioned later in this section), and is therefore less than the fall speed of most airborne organisms. **Wind speed** is usually expressed as metres a second, kilometres an hour, miles an hour, or knots (1 knot is 1 nautical mile an hour, or about 0.5 metres a second), and is measured by means of several types of **anemometer.** Wind speed near the ground can be estimated from its ability to move, for example, vegetation or the sea surface. Horizontal **wind direction** is usually expressed in degrees from true north (for example, an *east* wind blows *from* the east, 90°), and is measured by means of a **wind vane**; it can also be estimated simply by facing into the wind (as shown, for example, by a flag) and by using a compass.

Winds are caused by differential heating or cooling of the earth's surface, although often modified by barrier effects due to the earth's topography. Density differences develop on various space scales, and these in turn lead to pressure differences that may be depicted on weather maps by patterns of **isobars,** which are lines drawn on maps, joining places with the same pressure — corrected to some standard height, such as mean sea level. Air tends to flow from high to low pressure, but the earth's spin turns the flow *along* the isobars, not *across* them. Time is needed to bring about this turning. Hence, winds newly set up by differential heating or cooling will at first blow from high to low pressure. Examples of such winds are coastal and rainstorm outflow winds — see page 50. After some 5 to 10 h (longer nearer the equator), winds blow about parallel to the isobars, although that is not necessarily so when the isobar pattern itself is moving or changing shape. Steady wind flow along straight isobars parallel to the equator is said to be **geostrophic.** Wind systems seen on weather maps (page 68), usually lasting up to a few days, are more or less, though not quite, geostrophic — the differences are essential to their development, and to the formation of updraughts, clouds and rain. Hence, isobar patterns are a good guide to wind patterns, except at latitudes close to the equator, where the earth's spin about the local vertical is small and the turning effect is weak. These wind patterns often take the form of **vortices** and **waves** with dimensions of hundreds or thousands of kilometres — they are said to be on a **synoptic scale.** *Vortices* spin either in the same sense as that of the earth (when they are known as **cyclones**), or in the opposite sense (when they are known as **anticyclones**). A *wave* may be considered as a cyclone or anticyclone that is embedded in a broad, straight wind flow. **Global-scale wind systems,** such as the middle-latitude westerlies, the tropical easterlies (trade winds), and the monsoons, are also more or less geostrophic. In contrast, the **meso-scale wind systems** associated with coasts, mountains and rainstorms are shorter-lived and smaller. They are not geostrophic, unless the air takes a time of order 10 h to pass through them, as may happen with winds blowing across a highland area hundreds of kilometres wide.

Winds are always accelerating or decelerating on various time and space scales, and as a result they always either **converge** or **diverge.** Because the resulting air density changes are small, convergence and divergence lead to up and down motions. Convergence near the ground brings about an upward motion, and divergence a downward motion. These motions can lead not only to formation and dispersal of clouds and rain (page 40) but also to dispersion or concentration of airborne organisms (page 184). Convergence near the ground can be markedly strong near atmospheric fronts (page 52), and markedly long-lived in the inter-tropical convergence zone (ITCZ), which girdles the world near the equator, where tropical east winds meet from the northern and southern hemisphere, or where the easterlies meet monsoon west winds.

The speed and direction of the wind both change with height; hence surface wind need not be representative of the wind in which an organism is airborne. The two main causes for this vertical **wind shear** are friction due to ground roughness, and horizontal temperature gradients. Because the effect of ground roughness decreases upwards, the depth of the layer within which the wind is frictionally slowed is limited (Fig. 15a); it is called the **planetary boundary layer**, and it is often 500–1000 m deep. Wind *speed* increases with height within it more or less logarithmically, depending on roughness and lapse rate. Atmospheric mixing is inhibited by a small lapse rate, and the boundary layer is then shallow. When there is a temperature inversion, large vertical wind shears can occur. For example, on a clear, quiet night almost stagnant air near the ground can be overlain by a smooth wind of several m/s above heights of only a few metres. Wind *direction* also changes upward through the boundary layer — often clockwise in the northern hemisphere and anticlockwise in the southern, the difference being caused by the opposite spins about the local vertical. The change is often about 30°, but it can be more when there is a small lapse rate or an inversion. When the wind blows across a change in ground roughness, a new planetary boundary layer starts to grow upward from the ground. This can lead to a complex wind profile.

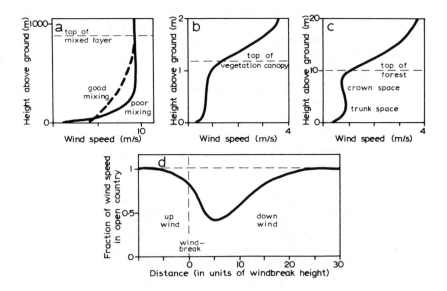

Fig. 15 – Typical wind speed profiles.
   a   Vertically, in open country, showing effects of good and poor mixing.
   b   Vertically, in dense vegetation, such as a field crop.
   c   Vertically, in a forest, showing a weak speed *maximum* in the trunk space (open canopy) and a weak *minimum* in the crown space (closed canopy).
   d   Horizontally, up wind and down wind from a porous windbreak — the curve shape varies with porosity.

A plant canopy is bent somewhat by the wind, and its roughness is therefore less than might be expected. Within a canopy, the wind profile is affected by **leaf density** (leaf area on unit ground area) and by the variation of leaf density with height (Fig. 15b). For example, in a forest, where tree crowns are much denser than trunk space, winds are much lighter below than above the canopy, but there tends to be a weak maximum in the lower trunk space (Fig. 15c). Wind entering the edge of a crop area is slowed by friction, but on leaving the downwind edge it is speeded up. Wind passing through a porous wind-break, such as a row of trees or an open fence, is also slowed by friction (Fig. 15d), particularly on the leeward side. This slowing affects the take-off and landing of organisms, not only directly but also through changing the heat and water balances of soil and vegetation.

Where winds are not geostrophic, vertical shears can be complex – as in coastal and rainstorm winds (pages 53 and 50). In contrast, where they are more or less geostrophic, wind shear between top and bottom of any layer in the atmosphere is more or less parallel to the **isotherms,** which are lines drawn on maps, joining places with the same temperature – in this case averaged through the depth of the layer. For example, with east-west isotherms and cold air on the poleward side (as occurs over most of the world for much of the time), there is westerly wind shear. Hence, the surface westerly winds typical of middle latitudes increase with height (Fig. 16a). Shears up to 5 m/s in a kilometre are common, and can be as much as 20 m/s in a kilometre. This can sometimes lead to strong winds at a height of only a few hundred metres when there is a near-calm close to the ground, and to winds stronger than 50 m/s at heights of 5–10 km. These **jet streams** are mostly associated with atmospheric fronts (page 52),

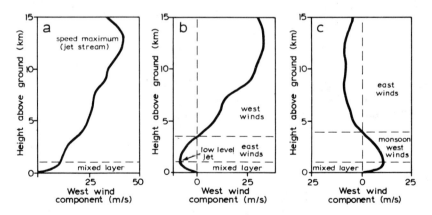

Fig. 16 – Typical wind speed profiles showing vertical shear above the planetary boundary layer (mixed layer).
    a   Westerly shear in middle latitudes.
    b   Westerly shear in low latitudes with surface east wind (trade wind).
    c   Easterly shear in low latitudes with surface west wind (monsoon).

where horizontal temperature gradients can be large. With westerly shear, surface tropical east winds are over-ridden by west winds at heights above a few kilometres (Fig. 16b). Ground roughness then leads to a speed maximum in the top of the boundary, a type of **low-level jet stream**; below the height where maximum speeds occur, frictional shear is great enough to outbalance the shear due to the poleward decrease in temperature. In monsoon winds, where temperature decreases towards the equator, and not towards the poles, there is *easterly* wind shear. Surface west winds are then over-ridden by east winds at heights above a few kilometers (Fig. 16c).

1.2.4 Rain

Rain, snow and hail fall from clouds, which themselves are formed mostly by condensation in moist air that is rising and is therefore cooling because it is expanding. From their shapes, clouds are of two main kinds: *heaped* or **cumulus** clouds, and *layered* or **stratus** clouds, according to their manner of formation. Convective updraughts tend to be localised and vigorous (up to about 10 m/s), and they lead to more or less detached cumulus clouds that individually change quickly (Fig. 17a). Other kinds of updraughts are often widespread and gentle (of order 10 cm/s), and they lead to extensive stratus clouds that change slowly (Fig. 17b). Sometimes convection develops only *after* layered cloud has formed resulting in the sprouting of convective towers from a layer. These clouds are

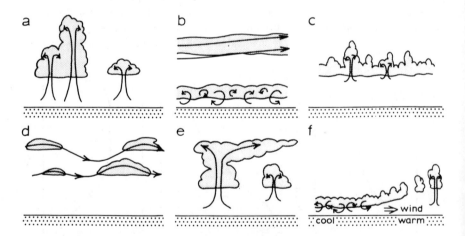

Fig. 17 – Cloud formation by the lifting of moist air.
a   Cumulus by convection over warm ground
b   Stratus, by mixing in the planetary boundary layer (below), and by slow but widespread lifting (above).
c   Castellatus, by convection from within other cloud.
d   Wave clouds, in the crests of gravity waves.
e   Spreading cumulus tops.
f   Stratus breaking into cumulus as it blows from cool to warm ground.

called **castellatus** (Fig. 17c). **Gravity waves** in the atmosphere may have clouds in their crests, and can lead to a complex patchiness within already existing stratus clouds (Fig. 17d). Lifting of the wind as it blows over mountains can produce clouds that more or less reflect the underlying topography.

It is common to have several kinds of clouds together, and for one kind to change into another. For example, cumulus tops can spread and combine into a layer (Fig. 17e); and boundary layer stratus clouds can break up and change into culmulus as a result of surface heating (Fig. 17f).

Satellite pictures show much detail about the spatial distribution of clouds, especially in relation to topography and the nature of the underlying earth's surface. Pictures taken at frequent intervals enable the growth and decline of cloud systems to be followed. Many clouds are related to atmospheric fronts and cyclones, and their patchiness reflects not only the updraught patterns but also the distribution of humidity caused by the varied histories of different parts of the cloud systems.

Clouds consist of minute water droplets or ice crystals, or of both, with sizes up to 10 μm and number densities of order 100/cm^3. These sizes and densities are determined by the water content, which is seldom more than 1 g/m^3, and by the number density of the specks, or nuclei, on which water vapour condenses. Cloud particles fall through the air at speeds less than 1 cm/s, which is much less than most updraughts in the atmosphere. This is why clouds appear to float in the air.

Few clouds give rain. In those that do, some of the cloud particles grow large enough to fall out and perhaps reach the ground. Growth is either by coalescence as a result of collision, or by distillation of vapour from supercooled droplets to crystals. Collision of droplets alone leads to rain and drizzle drops; collision of supercooled droplets with ice crystals leads to hail stones; collision of ice crystals leads to snow flakes. Distillation enables a few crystals to grow large at the expense of many supercooled droplets. Snow and hail can melt to rain before reaching the ground. Growth takes about one hour – or less in clouds that have a large water content. Thus, for a cloud to give rain it must last for about an hour or more; it must also be deep enough that the growing particles do not fall out too quickly – about 0.5 km for stratus and 2 km for cumulus.

Cumulus and **castellatus** clouds give local, intense and short-lived rains called **showers**. They are *local* because the convective updraughts have widths comparable with their depth; *intense* because the condensed water is at first stored in the rapidly growing and dense cloud, and then released (at rates commonly 10-100 mm/h) when the drops have grown large enough to fall against the updraught; *short-lived* because rain-induced downdraughts tend to replace the updraughts. **Stratus** clouds give more widespread, lighter, but persistent rains. They are *widespread* because the cloud layers are extensive; *lighter* because the water falls out at about the same rate as it condenses from the vapour (at rates of order 1 mm/h); *persistent* because the cloud-forming updraughts last hours and sometimes days, and the rainy cloud sheets may take hours to pass overhead. Deep-cumulus rains are typical of low latitudes, and deep-stratus rains of high latitudes.

The fall speeds of raindrops increase with size: a 0.1 mm diameter drop falls at 25 cm/s, a 1 mm drop at 4 m/s, and a 5 mm drop at 9 m/s. Drops larger than 5 mm diameter are unstable and break up during fall.

Rainfall rate or **intensity** is the rate of increase in depth of liquid water on a horizontal surface, excluding run-off but including melted snow and hail. It is usually expressed in mm/h, and measured by various kinds of **raingauge**. Rates are very variable, but seldom exceed 100 mm/h. The most intense falls are usually of short duration — very seldom heavier than 10 mm/min. Drops falling at any given time usually have a wide range (spectrum) of sizes that is easily seen by the splashes on a dry surface briefly exposed to falling rain. Intense rains usually fall mostly as large drops. Although there may also be many small drops falling at the same time their contribution to the rainfall rate is small. Sometimes there is a remarkable uniformity of drop size. Examples of this are where rain falls from shallow cloud (where the drops are usually small — say, less than 0.5 mm diameter); or where it falls from cloud whose base is high enough for the smaller drops to evaporate before they reach the ground, and where those that do are usually of medium size (say, 1-2 mm diameter).

Splash rate is the number of drops striking a horizontal surface of unit area in unit time. It varies with rainfall intensity and drop size. For an intensity of 1 mm/h and drop diameter 1 mm (that is, drop volume about 0,5 mm^3), the average splash rate is about $2/mm^2$ each hour. Doubling the intensity at fixed drop size doubles the splash rate, but halving the drop size at fixed intensity increases the splash rate 8-fold. Thus, the average splash rate for a 10 mm/h rainfall with 0.5 mm diameter drops is about $150/mm^2$ each hour.

1.3 DIURNAL WEATHER CHANGES

Weather changes between day and night (called **diurnal changes** by meteorologists) are well known. Warm, dry and windy days often contrast with cool, damp and still nights. Temperature will be considered first because diurnal variations of all other weather elements depend strongly on that of temperature.

1.3.1 Temperature

On windless days, daytime warmth and night-time coolness are a result of *local* heating and cooling of the air, whereas on days with wind there is the possibility of bringing in warm or cool air from elsewhere. Consideration of the latter changes is left until later sections in this Chapter.

Ground surface temperature varies from a maximum in the early afternoon to a minimum around dawn (Fig. 18). At these times the upward flow of heat

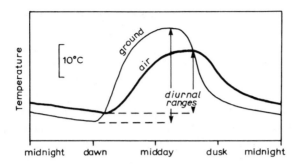

Fig. 18 — Typical day-to-night variation of air and ground temperature in clear, quiet weather.

into the air just balances the downward flow. Both size of the **diurnal range** of temperature, and the shape of the temperature curve during the day, vary greatly: they depend on *season*, latitude and cloudiness (affecting sunshine intensity and duration), on the *nature of the ground* (slope, aspect, density, porosity, composition — affecting reflectivity and the storage and downward flow of heat, and surface evaporation of water), and on *temperature, humidity and speed of the air* (affecting upward flow of heat). Seas, lakes and large rivers have diurnal ranges of less than 1°C whereas, by contrast, diurnal ranges of bare, dry soil can exceed 50°C. Because the *air* is largely heated and cooled by contact with the ground, the diurnal range of air temperature decreases upwards, becoming only about 1-2°C above the convective layer (Fig. 14a). At the standard measuring height for 'surface' air temperature (about 1 m above the ground), diurnal range is commonly 10-20°C in the warmer months over land, but almost nil over the sea, and even over land in cloudy, windy weather of the cooler months in middle and high latitudes. Maximum *air* temperature tends to be around mid afternoon, when ground temperature is already falling, and the minimum is about dawn (Fig. 18). Depth of the convective layer also has a corresponding diurnal variation — from nil about dawn to a maximum about mid afternoon (Fig. 14a).

In a *valley*, shading from the sun as well as long-wave radiation from the sides can both add to the complex exchange of heat. Among *vegetation*, each leaf and stem has its own diurnal range, due to its particular orientation towards sunshine, its exposure to the wind, and its shading by other leaves and stems. In dense vegetation, such as a field crop or a forest, the greatest diurnal range is near the canopy top, and air temperature may be almost constant near the ground (Fig. 14b).

1.3.2 Humidity

In the absence of *horizontal* advection, vapour pressure can change only by evaporation or condensation, or by *vertical* advection. It is these processes that cause the diurnal variation of vapour pressure (or dew point). Evaporation is possible if the air is unsaturated, but even then it will happen only if the soil is moist or the vegetation is transpiring. Where there is such a source of water vapour, evaporation rate grows during the morning as temperature rises and saturation deficit increases. At the same time, mixing takes place through a deepening convective layer. If there is already a hydrolapse, the less humid air aloft is mixed downward; as a result, near the ground there tends to be a weak humidity maximum during the morning and a minimum in the afternoon. Continued evaporation, but a rapid decline in convection, then leads to a rise of humidity into the evening because the newly added water vapour tends to stay in the lowest few tens of metres of the atmosphere. If, later, the temperature of the ground or vegetation falls below the dew point, then dew or fog will form, and the resulting condensation will lower both vapour pressure and dew point. Thus, there tend to be two weak vapour pressure maxima, about mid morning and during the night, and two weak minima, mid afternoon and about dawn — see Fig. 19. The size of the **diurnal range** of vapour pressure will depend on the availability of water and heat for evaporation, the depth of atmospheric mixing, and the hydrolapse before mixing starts. By contrast, the diurnal variation of *relative* humidity, although dependent on changes in both vapour pressure and temperature, usually has only one maximum and one minimum, because temperature effects on saturation vapour pressure outweigh the effects of evaporation and condensation on actual vapour pressure. Thus, from the time of temperature minimum (about dawn) to that of temperature maximum (mid afternoon), relative humidity falls, whereas for the rest of the 24 hours it rises (Fig. 19).

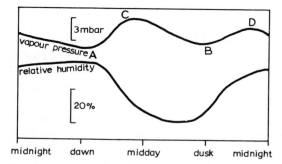

Fig. 19 – Typical day-to-night variation of air humidity in clear, quiet weather. Water vapour pressure has two *minima*:

    A, at the time of minimum air temperature,
    B, when upward flow of water vapour from the ground becomes greater than the upward flow by mixing the drier air aloft.
And two *maxima*:
    C, when upward flow of vapour from the ground becomes less than the upward flow by mixing with drier air aloft.
    D, when dew starts to form.

1.3.3 Wind

Winds near the ground tend to be slowed down by ground roughness. This slowing effect is greatest nearest the ground, so the wind strengthens with height, and there is a **wind shear**. When there is convective stirring of the atmosphere, faster moving air from aloft is carried downward and wind shear is therefore weak; but when there is a temperature inversion, with little or no stirring, wind shear is strong, and hence near the ground the wind can fall to calm. Drag due to ground roughness is therefore most effective in slowing down the wind during cloudless weather, when the diurnal variation of lapse rate is large. At such times the wind is strongest in the afternoon and weakest at night (Fig. 20). In clear but windy weather, although speeds still tend to be least at night, the proportional change between day and night is not as great as in quiet weather. Such a *mid afternoon* maximum is characteristic of middle latitudes, where the westerlies often increase with height above the planetary boundary layer. By contrast, in tropical easterlies and monsoons, where speed usually decreases upwards above the planetary boundary layer, the maximum tends to be in *mid morning*, for deepening of the convective layer later brings down *slower* air from aloft (Fig. 20).

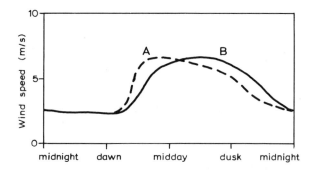

Fig. 20 – Typical day-to-night variation of wind speed in clear, quiet weather.
A   Tropical east winds and monsoon westerlies.
B   Middle latitude westerlies.

1.3.4 Rain

Diurnal variation of rainfall incidence and amount depends upon the diurnal variation of cloud depth and persistence.

Cumulus clouds *over land,* formed by sunshine heating the ground, have a characteristic daily cycle of development. They appear when the convective layer has deepened enough to allow condensation to occur in the rising thermals (page 49). Because the height of cloud base increases with dryness of surface air, cumulus clouds first appear late in the morning if the air is very dry, or they may not appear at all. Large clouds may last an hour or more, smaller ones much less, 'so although the sky may not appear to change as a whole, individual clouds constantly come and go. Latent heat released by condensation adds to the buoyancy of cloudy thermals, so that some clouds may tower up well above the top of the convective layer. Thus, both base and top rise from first formation until late afternoon, when convection begins to die away. Hence cumulus clouds over land tend to be largest and most persistent in late afternoon, which is why showers are heaviest and most frequent at this time of day. Showers also tend to form earlier over mountains than over valleys and plains. Convection clouds *propagating through* the atmosphere, having started at some preferred site, may lead to peak shower frequency being later in the day than expected. Large castellatus clouds can be triggered after dusk by sea breeze fronts and by upslope winds, the latter probably accounting for night-time rainfall maxima over large tropical highlands. Sea breezes can lead to a morning rainfall maximum at the coast. *Over the sea* there is little diurnal variation of cumulus clouds; surface temperature and humidity there, and hence cloud base, change little.

Boundary-layer stratus clouds are seldom deep enough for growth of drops larger than those in drizzle. Places where such clouds are frequent, for example coasts with onshore winds that have blown across cold water, can have a well-marked diurnal variation of drizzle, with a maximum in the early morning. This is because stratus clouds in the planetary boundary layer over land have a characteristic daily cycle. Daytime heating is usually enough to set up shallow convection, and hence cloud base rises as surface relative humidity decreases. When the base of a cloud rises to the same height as its top, the cloud may disperse, usually by mixing with dry air from aloft. Sometimes the breaking stratus cloud changes to cumulus. Higher stratus clouds, for example those that occur at fronts or in cyclones, show little diurnal variation of either amount or height. Rain from higher stratus clouds therefore also has little diurnal variation. Stronger variations occur in stratus clouds formed by windflow over mountains.

1.4 SUDDEN WEATHER CHANGES

All weather changes are in the first place caused by heating and cooling of the atmosphere, as can perhaps be most clearly seen in the diurnal changes, discussed in section 1.3. Some changes are more sudden: we can all recall the onset of a big wind, a heavy rain, or a temperature fall that took place within minutes. Such changes cannot be due to a correspondingly sudden heating or cooling of the atmosphere. Instead, the changes are due to *advection*, that is to say a windborne inflow, from elsewhere, of the rain, low temperature or whatever. One airstream pushes away another and takes its place; in particular, cool air undercuts warm. To understand a sudden weather change we therefore need to understand sudden *wind* changes. It is convenient to divide the following discussion into three sections according to the horizontal scale of the wind systems involved: **gusts,** up to about one kilometre; **squalls,** up to some tens of kilometres; and **fronts,** up to some hundreds of kilometres.

1.4.1 Gusts

Changes of wind speed and direction on time scales of a few seconds can be readily felt on most days, but they are dramatically revealed by continuously recording anemometers and wind vanes. Fig. 21 is an example of such a record. The amount of detail varies with the sensitivity of both the measuring and recording parts of the instrument, but the main characteristics are clear: never-ending and irregular fluctuations. The sudden increases in speed are **gusts,** and the decreases are **lulls.** Gusts and lulls are of great importance in affecting not only the take-off of organisms (page 22) but also their windborne spread (Chapter 3) and their landing (Chapter 2). They are caused by a succession of innumerable passing atmospheric **eddies,** which are of two kinds: those due to the *roughness* of the ground, and those due to *buoyancy* of some parts of the air.

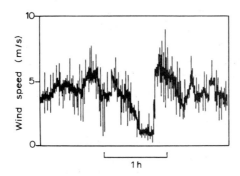

Fig. 21 – Example of a strip chart from a wind speed recorder (anemograph) showing:
   brief gusts and lulls,
   irregular pulses lasting a fraction of an hour,
   longer, and even sudden, changes.

Roughness eddies resemble the turbulence that can be seen in a river flowing over its stony bed, where there is an irregular array of constantly changing and interacting swirls, travelling mostly downstream. Because the ground is rough, some parts stand up against the wind and slow down patches or streaks of air. Elsewhere, the wind blows over and around these barriers to form a variety of eddies, some of which are shed and move downwind and in turn become broken by other barriers. These barriers vary in size from smaller than soil particles, through leaves, whole plants and buildings, to hills and mountains. Eddies can be seen in drifting snow, dust, leaves or smoke. Many of them can be looked upon as being spherical or cylindrical masses of air, tumbling or rolling forward about more or less *horizontal* axes; others turn about *vertical* axes. As these eddies pass by they lead to gusts and lulls. In the extreme, lulls are near calm and gusts have about twice the wind speed averaged over periods of, say, minutes. More often, lulls are about 0.5, and gusts about 1.5, times the mean wind speed. At night, especially in light winds, fluctuations are weak and can be less than 0.25 of the mean. This difference between day and night reflects the difference in ease of mixing in the vertical. On a warm, sunny day, and even at night in strong winds, mixing is vigorous and fluctuations are large. By contrast, on a clear night with light winds, when air becomes cold near the ground and there is a temperature inversion (page 33), mixing is weak and fluctuations are small. When considering the effects of wind speed on take-off (page 19), it is therefore necessary to bear in mind that there will be gusts and lulls with speeds different from that of the mean wind, and it is these sudden changes from one to the other, and not just the strength of the *mean* wind, that affect take-off.

Another result of mixing by eddies is the slowing of the wind near the ground, whilst aloft it continues to blow more or less unaffected. Such a vertical **wind shear** has to be borne in mind when considering the speed of windborne organisms aloft in relation to winds measured only near the ground. Vertical wind shear tends to be large when mixing is weak, but small when mixing is strong (Fig. 15a, see also Fig. 128). The frictionally slowed layer is known as the **planetary boundary layer** (see page 38). Its thickness limits the height to which airborne organisms can be carried upward by roughness eddies. The largest, and the rarest, eddies have a size comparable with the depth of the boundary layer; they lead to the least frequent gusts and lulls (mean intervals of a few minutes). Small eddies are numerous and lead to gusts and lulls with mean intervals of seconds or less. Vertical wind shear is *unstable*; this is because the kinetic energy of a sheared flow becomes less after mixing. If enough energy is released, eddies can be raised or lowered even when such movement is resisted by the presence of a lapse rate (page 33) less than about 1°C in 100 m. The smaller is the lapse rate, the stronger is the shear that can persist without breaking down into eddies. Close to the ground, eddy size tends to be smaller. Within a few millimetres, however, eddies of that size and smaller are strongly suppressed by internal friction, known as **air viscosity**. Hence, close to the ground there is a shallow film of non-gusty air, known as the **viscous boundary layer**. Such layers develop over all surfaces exposed to the wind, including, for example, both upper and lower surfaces of leaves. Airborne particles there cannot be carried into the turbulent air nearby; instead, their vertical motion is controlled by gravity, and by the airflow around the barrier over which the viscous boundary layer has developed.

Buoyancy eddies resemble the motion within water that is being heated from below — where masses or columns of warm, less dense and therefore buoyant water rise through their cooler surroundings, mixing as they go. When the earth's surface is warmer than the air, some parts are warmer than others and there develops a patchiness of air temperature, especially in the lowest few tens of metres of the atmosphere. Each warmer patch tends to rise and is replaced by sinking and inward-flowing, cooler air. Over-turning circulations develop, with rising air surrounded by sinking air. The rising warm air mixes with its cooler surroundings, thereby becoming less buoyant, but later it spreads sideways at some height where it loses its buoyancy altogether because it enters surroundings that are as warm as itself. A circulation as a whole may become detached from the ground and rise as an overturning, dome-shaped mass of air known as a **thermal**; or a **column** may persist with its base near the ground and its top feeding a slower-rising thermal (Fig. 22). Similar, but stronger, circulations are easily seen in the smoke from fires and explosions. On a sunny day, the convective layer (page 33) becomes filled with buoyant eddies or their remains; some of them reach the top of the layer and may carry airborne organisms to that height, sometimes at speeds as great as several metres a second. As these buoyant eddies drift by, they cause fitful fluctuations of the wind, more irregular than those due to ground roughness, and often superimposed upon them. Sometimes a rising column rotates and becomes concentrated into a strong, narrow **whirl,** or 'dust devil', with wind speeds greater then 10 m/s over radii up to a few tens of metres, able to pick up loose, light objects from the ground, such as dust and dried leaves, and carry them towards the top of the convective layer, along with any airborne organisms (Fig. 22). Sometimes convective eddies are arranged in **rows** along the wind shear direction, or in a **cellular pattern** consisting of polygonal rings with either sinking air at the middle and rising air at the edges, or the other way round. Row spacing and cell width are several times the depth of the convective layer (Fig. 22). The kinds of eddies present at any given time will depend on ground roughness, topography, vertical shear and lapse rate.

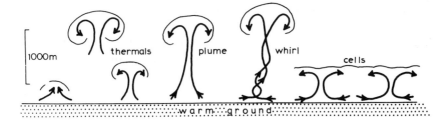

Fig. 22 – Various kinds of convective elements in air over warm ground.

Wind gustiness is accompanied by corresponding fluctuations in air temperature. These fluctuations are small — seldom more than 1-2°C — especially those that are due to roughness eddies, whose continuous mixing tends to smooth out irregularities. Temperature fluctuations are greatest where the lapse rate (page 33) differs markedly from 1°C in 100 m. If the lapse rate is larger, as can happen close to the earth's surface when it is much warmer than the air, both kinds of eddies bring down patches of cooler air and carry up warmer air. By contrast, if the lapse rate is less than 1°C in 100 m, eddies carry heat in the opposite direction; and the overturning, being against gravity, consumes energy and becomes more difficult with smaller lapse rates. Gustiness then tends to be damped. A sudden increase in wind speed at night enhances mixing, increases the lapse rate and raises the air temperature near the ground.

Over country having a patchwork of vegetation, soil types and topography, perhaps on a variety of space scales, a complex pattern of temperature can be set up, day or night, when there is little wind. Fluctuations of up to a few °C and lasting a fraction of an hour, can then occur. Strong winds tend to smooth out such irregularities.

As with temperature, small but rapid fluctuations of humidity are typical of sunny days: changes of dew point by 1°-2°C can take place within a minute when there are wind gusts in the presence of a hydrolapse (page 36). With the usual upward decrease of dew point, minor jumps in humidity occur with upward-moving gusts, and dips with downward-moving gusts. In contrast, when humidity increases upwards, a sudden increase in wind can bring down to the ground air with higher humidity.

1.4.2 Squalls

On some days there can be a sudden strong wind lasting as long as a few tens of minutes. Such a wind is known as a **squall**. Both the mean wind and gusts during a squall can be several times as strong as before the squall. Moreover, at the onset of the squall there may be a sudden change of direction. A squall can suddenly carry spores, pollen and seeds into the air, as well as cause drifting of dry soil particles along with their accompanying organisms; it can also sometimes induce flying insects to settle. The changed wind direction may well bring in different kinds and numbers of windborne organisms.

Most squalls are the products of rainstorms. Drops falling in heavy rain drag down the surrounding air and start a downdraught. Below cloud base there is some cooling of the downdraught by partial evaporation of the drops, and the resulting negative buoyancy adds to the downward motion. When the cooled air strikes the ground it spreads outward on a more or less circular front (Fig. 23a). Peak gusts may be 20-50 km/h, and even 100 km/h in hot, dry weather, when much damage may be caused. Where a rainstorm is moving across country, the squall tends to be strongest on the downwind side, and weakest (or even absent) on the upwind side. In a cluster of rainstorms, each with its own life cycle, the downdraughts interact and so lead to very irregular winds – typical of thundery weather (Fig. 23b). Sometimes, where rainstorms form more or less in line, with their downdraughts combining into a **line squall,** they may cross country at a speed of 50 km/h or more (Fig. 23c). Downdraughts from a large area of rainstorms can spread out for tens or hundreds of kilometres. With the wind change there is a temperature fall (often 5°C, sometimes 10°C), and usually a relative humidity rise, although not necessarily a rise in dew point, as well as an increase in gustiness, all of which can lead to changes in take-off rate.

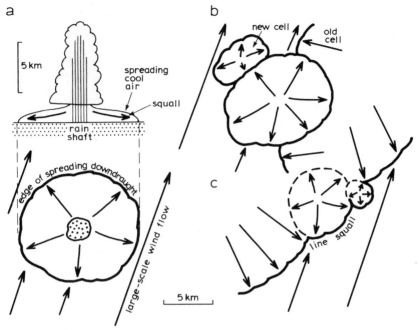

Fig. 23 – Squalls.
  a   Section and plan views of a cool downdraught spreading outwards from a
      rain storm.
  b   Interaction of outflows having various ages.
  c   Outflows from a line of storms combining to form a line squall.
Winds are shown *relative to the storm centres,* which are themselves moving.

The cool air of a downdraught squall undercuts and lifts the warmer air
ahead of it, along with any airborne organisms. Fig. 24 shows schematically
the windflow relative to the leading edge of the squall. Flying insects may react
to such lifting, which would otherwise tend to carry them to heights too cold
for flapping flight. Descent would then take them into the squall and hence

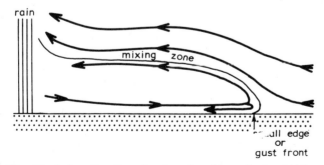

Fig. 24 – Vertical section through a downdraught squall, showing flow *relative
to the leading edge,* which is itself moving away from the rain storm.

back towards the leading edge, where they would tend to gather. This may be a mechanism for concentrating flying insects – the squall would act as a kind of broom sweeping through insect-laden air (see page 197).

Squalls also occur near mountains, even in the absence of rainstorms, and there they are caused by the fitful growth and breakdown of eddies in the wind as it blows over and around the mountains, particularly when there is a temperature inversion at some height near the mountain tops (Fig. 25). Complex spillage patterns can then appear like those in a river flowing over its rocky bed. Downdraught squalls from rainstorms also become blocked and diverted by mountains, and the resulting complex and rapidly changing wind patterns are not only difficult to record but also impossible to forecast in detail. Rainstorms among mountains can be expected to lead to correspondingly complex patterns of take-off, movement and landing of windborne organisms.

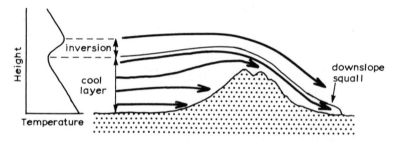

Fig. 25 – Vertical section through a downslope squall, showing the cool layer deepening and spilling over a mountain range (and through passes).

1.4.3 Fronts

Whereas the wind after a squall often goes back to its former direction within an hour or two, there are days when a sudden change in direction is maintained for many hours, or even a few days. Such a change accompanies the passing of a **windshift line** or **front** that marks the junction between two airstreams with dimensions of hundreds or thousands of kilometres. A front is so large that it can seldom be made visible as a whole to a single person – unlike thermals, dust devils are downdraught squalls, for example. Instead, its existence is revealed by weather maps, based on the simultaneous observations of many weather watchers (see page 68), or by pictures of associated cloud patterns taken from aircraft or earth satellites.

The passing of a front can have a marked effect on the rate of take-off by organisms. The changes in temperature and humidity, particularly when modified by cloudiness and rainfall, can alter the release rate of spores, pollen and seeds, for example, and changes in wind speed and gustiness affect take-off and landing by insects.

Fronts are associated with horizontal temperature gradients. The leading edge of a downdraught squall is also a front, but here we are concerned with fronts mostly on a larger scale. For convenience, they can be divided into three kinds according to the causes of the temperature gradients, which are brought about by differential heating by the sun of:

> land and sea
> mountains and plains
> low and high latitudes.

Along a coast in warm, sunny and quiet weather, an onshore wind can often be felt from about mid morning onwards, becoming strongest by the afternoon, and dying away in the evening. Such a wind is called a **sea breeze**; it may exceed 15 km/h by mid afternoon. In low latitudes, sea breezes are very common; in middle and high latitudes they are largely confined to the warmest months, but seldom blow on cloudy or windy days. Their mechanism is straight-forward. Although air temperatures over land rise during the morning, over the sea there is little change, and so a density difference is set up between land and sea. Cool, dense sea air starts to flow inland, and a circulation is set up with a seaward flow aloft. The sea breeze deepens to 500-1000 m by mid afternoon with greatest speed at about 200 m (due to ground roughness), and the seaward flow aloft may be just as deep (Fig. 26). The forward edge of the sea breeze is a **sea breeze front** that spreads inland with a speed of 5-10 km/h; it may reach 50 km from the coast by sunset, and even 100 km or more by mid evening,

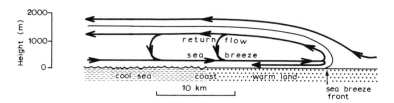

Fig. 26 – Vertical section through a midday sea breeze, showing flow *relative to the sea breeze front*, which itself is moving inland. (Note the similarity to the downdraught squall, Fig. 24.)

especially in hot countries. The front moves slower than the breeze, until late afternoon, and it has a narrow band of rising sea air a few hundred metres across, with updraughts of about 1-2 m/s that help feed the seaward return flow aloft. As the sea breeze front passes overhead, the wind suddenly starts to blow more or less from the coast. Not only is there often a fall in temperature of several °C (and perhaps a comparable rise in dew point) in as many minutes, but also the kinds of airborne organisms passing overhead change from those typical of land to those typical of sea or of the strip of land between the coast and the sea breeze front. Some organisms, however, may fall or fly from the seaward-flowing part of the circulation aloft into the sea breeze beneath, and so lead to a mixture of organisms in the sea air. A sea breeze front behaves in this way like the leading edge of a downdraught squall from a rainstorm (page 50), and flying insects may become concentrated there. Trapping there may prevent them from being taken out to sea by day, and organisms carried in an offshore wind that was blowing before the sea breeze started may be carried onshore again.

On a quiet, cloudless night, an opposite coastal wind may develop as land air becomes colder than sea air. Such a **land breeze** is usually weak and shallow: less than 5 km/h, and at most only a few hundred metres deep. Windborne organisms can be carried some tens of kilometres out to sea by dawn. Subsequent onset of a sea breeze can then bring the air back over land, so the sea breeze air need not be devoid of organisms that come from the land. Along hilly coasts, sea and land breezes are made more complex by channelling along valleys and spillage over ridges.

A similar kind of breeze can develop over a city that is warmer than its surrounding countryside. City warmth, known as the **urban heat island,** can be due to differences in heat and moisture balances that exist between city and country, caused by differences in the properties of the ground, or by the presence in the city of artificial heat sources.

Over a hill slope in warm, sunny and quiet weather, an **upslope** (or **anabatic)** **wind** can sometimes be felt from about mid morning onwards. Like the sea breeze, this wind dies away soon after dusk, and even earlier on short or eastward-facing slopes. The strength of an upslope wind depends on such properties of the slope as its inclination, aspect, length and vegetation cover, but it is seldom stronger than 5 m/s. Along a ridge crest, upslope winds from opposite sides can meet at a more or less stationary windshift line, where the slope winds feed an updraught that forms part of a circulation with downdraughts over the valleys. Organisms in valley air can then be taken up to the ridge crest and higher, where the upslope winds help buoyant gusts to mix the convective layer of the atmosphere (page 33). Indeed, the convectively mixed layer tends to be deepest over the highest ground, partly as a result of lifting as the layer passes over a ridge. In a valley closed at one end, upslope winds induce **upvalley wind** along the valley floor.

At night, a **downslope** (or **katabatic) wind** tends to form as the air cools, becomes dense and begins to flow downhill. Downslope winds are weak, as is shown by smoke drift, but when coming from opposite sides of a valley they may combine at the bottom to give a cool, **downvalley wind.** In a steep, narrow valley, speeds can reach 5 m/s, and the leading edge becomes like that of the cool outflow from a rainstorm. At the valley mouth, where it opens on to a plain, a strong downvalley wind may persist as a **jet** up to a few hundred metres deep and with a breadth like that of the valley. This jet may channel fungal spores, for example, across a narrow part of the plain (Fig. 27). In mountainous country, a complex pattern of upslope winds by day and downslope winds by night is not uncommon in clear, quiet weather, and this can lead to a correspondingly complex pattern of the take-off, movement and landing of windborne organisms. In cloudy or windy weather, such local breezes are much less likely to appear.

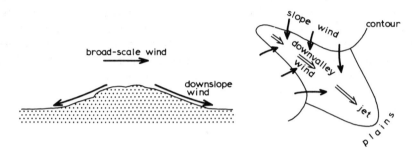

Fig. 27 – Downslope winds at night in clear, quiet weather. A jet may form over the plains, fed by a downvalley wind.

Updraughts that occur at these coastal and mountain fronts may set off the formation of rainstorms, and any resulting downdraughts then interact with the coastal and mountain breezes to produce a rapidly changing pattern of winds that can be detected satisfactorily only by means of a dense network of observations.

Whereas the sudden wind change following the passing of a coastal or mountain front usually lasts some hours, there are others that last up to a few days and are due to the passing of fronts or windshift lines on scales of hundreds or thousands of kilometres. They are caused by a concentration of the global-scale temperature increase from pole to equator into zones tens of kilometres wide, and hundreds or thousands long, across which the temperature may change by 10°C or more. This concentration is usually brought about by convergent winds within cyclones (page 37), and fronts can often be seen on weather maps for many parts of the world (page 68). The wind changes are again accompanied by changes in temperature and humidity, and often also by cloudiness and rain. These fronts are of two main kinds, according to their movement: **warm front,** where warm air replaces cold, and **cold front,** where cold air replaces warm. Temperature changes can be as large and rapid as those at coastal fronts and rainstorm outflows, but they are often smaller and slower and masked by the effects of differences in diurnal variation of cloudiness and rainfall from one side to another. For example, there may be boundary layer stratus cloud on one side and cumulus on the other; sunshine tends to disperse the former but promote the latter. Dew point usually changes as a front passes overhead, although not necessarily in the same sense as temperature. Hence, relative humidity may change rapidly or remain more or less unaltered, depending on the occasion. Some fronts are marked by big changes in both dew point and relative humidity, but little change in temperature; these are **dry lines**. Humidity changes at fronts are sometimes more complex due to evaporation of falling rain and differences in diurnal variation from one side to the other.

It has been seen that all kinds of fronts are accompanied by more or less sudden changes of wind speed or direction, or by both together — that is, there is *horizontal* wind shear as well as *vertical*. This complexity of structure is increased where effects are combined; for example, where downdraught squalls accompany a large-scale front that is crossing a mountainous coastline where topographic eddies and the sea breeze are already interacting.

CHAPTER 2

Staying airborne

The distance that most airborne organisms can move depends upon the time that they are airborne and the wind speed. For organisms that move *through* the air, however, such as insects in flapping flight, it may depend more on their own air speed. *Movements* of the air and of the organism through the air are discussed in Chapters 3–7; here we consider the *duration* of flight.

Every airborne organism tends to fall to the ground because it is denser than air. This **fall-out**, or **sedimentation**, is slowed down by updraughts and by friction; and it is speeded up by downdraughts, by wash-out in rain, and by capture on vegetation (Chamberlain 1967). In addition, flapping flight by insects affects the rates of both rise and sink; and this, too, is modified by atmospheric properties, notably temperature, cloudiness and humidity. Thus, the weather exerts complex and variable effects on the ways by which organisms leave the atmosphere, and hence on the time they spend airborne.

2.1 FALL-OUT

In still air, an organism will fall downward at a steady speed (the terminal velocity, settling speed, or **fall speed**) when its weight is just balanced by friction. Other properties being the same, spherical organisms fall fastest. For a density of 1 g/cm^3, the fall speed of spheres increases with size as follows:

radius (μm)	1	2	10	50
fall speed (cm/s)	0.01	0.3	1	20

Such spheres starting from rest reach these fall speeds in a small fraction of a second. Organisms with the same mass but much smaller density (and therefore larger surface area because they have air-filled spaces) are rather like air-filled balloons and fall more slowly. Some seeds and fruits behave in this way, and so do the pollen grains of *Pinus*. Organisms with given volume and density fall slower if they are flattened, and particularly if they are long and narrow. Hence, spherical organisms covered with spines or hairs also fall much slower than if the whole volume lay within a sphere. The curved spines on the uredospores of *Puccinia graminis* var. *tritici,* the cause of **wheat stem rust,** may have this effect (Orr & Tippets 1972). Most spores and pollen grains probably remain airborne up to only a few days, and usually much less, but such a time would often be enough for a drift of several hundreds or even thousands of kilometres (page 77). Some seeds, such as those of cotton, *Gossypium,* are covered with hairs; others have tufts (for example, *Salix, Populus*) or parachute-like structures (for example, the pappus of some Compositae), all of which reduce the fall speed. With first stage caterpillars of the **Douglas-fir tussock moth,** *Orgyia pseudotsugata,* as well as those of the **eastern spruce budworm moth,** *Choristoneura fumiferana,* both of which spin silk threads that tend to keep them airborne, laboratory studies show that fall speed decreases as thread length increases (Batzer 1968, Mitchell 1979, Fig. 28). The thread drifts somewhat ahead of the caterpillar, and when it becomes entangled with an obstruction the caterpillar whips about until it finally lands. Adult insects, with wings held straight or folded, have fall speeds up to a few metres a second. The fixed wings on some tree seeds reduce fall speed not only directly but also indirectly by generating some lift through spinning. Maple (*Acer*), ash (*Fraxinus*) and tulip tree (*Liriodendron*) behave in this way (McCutchen 1977).

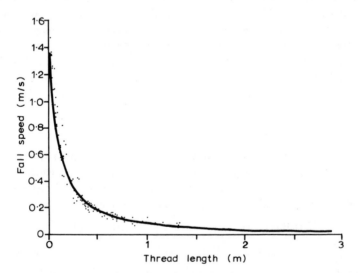

Fig. 28 – Variation of fall speed with silk thread length for 153 second-instar caterpillars of the eastern spruce budworm, *Choristoneura fumiferana.* (After Batzer 1968).

Fall speed is not affected directly by a *horizontal* wind. But all winds have updraughts and downdraughts, often strong when the atmosphere is being heated from below, but usually weak when there is a temperature inversion,

especially near the ground on clear nights. The effects of such opposite vertical
motions can be expected to cancel, so that *average* fall speed in a cloud of
organisms is little changed. Some individuals, however, will be more affected
by updraughts than by downdraughts during a given time interval, and others
more by downdraughts. The latter individuals will become trapped by vegetation
faster than if they had been falling through a horizontal wind, and so they
cannot be taken aloft again in any updraughts they would soon have met. Thus,
turbulence eddies help considerably to deplete the atmosphere of its load of
organisms through impaction (below), but not by affecting fall-out. There is no
clear evidence that larger-scale downdraughts, such as those associated with
fronts, and in showers and thunderstorms (page 50), affect the fall-out of airborne organisms. Deep and powerful updraughts can undoubtedly carry organisms upwards through many kilometres (leading, for example, to deposition in
mountain snow). Insects would often be killed by cold in this way, but spores
might be more resistant. Organisms drifting in the presence of more common,
but shallower, daytime convection can still encounter updraughts of order
1 m/s lasting some tens of minutes, and reaching a kilometre or more above the
ground. A small insect with fall speed around 1 m/s could then remain airborne
for, say, an hour whilst it rises and falls, with little need to use energy in flapping flight.

2.2 IMPACTION

Windborne organisms may bump into anything that stands out against the wind.
As the wind approaches an obstacle the flow is turned aside and it splits, joining
again downwind (Fig. 29). A windborne organism will also tend to be carried
around an obstacle, but because of its inertia it is turned aside less than the wind

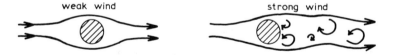

Fig. 29 – Schematic separation of wind flow around an obstacle: streamlines in
a weak wind; turbulent wake in a strong wind.

and so there is a chance of bumping, or **impaction**. Small organisms and light
winds are less likely to lead to bumping than large organisms and strong winds.
For given size and wind strength, the chance of bumping increases with obstacle
size. It is no wonder, then, that the spores of fungi that are stem and leaf pathogens of cereals tend to be larger than the spores of stigma and glume pathogens.
Smaller spores, such as those of soil fungi cannot be caught by impaction and
must therefore leave the air by some other mechanism. Tree trunks can catch
efficiently only large spores, such as those of the **lichen** *Pertusaria*, which have

150 μm diameter. Such spores, however, have large fall speeds and therefore do not drift as far as smaller spores. It may be that many stem and leaf fungal pathogens have spores with diameters of 10-30 μm as a compromise between fall-out and impaction.

Spores are less likely to bounce if the surfaces they strike are waxy, wet or hairy, or if the spores themselves are sticky, as can happen in moist air (Gregory 1971). Wind-tunnel experiments with ascospores of *Eutypa armeniacae,* a pathogen of apricot, *Prunus armeniaca,* showed that they were more efficiently caught by leaf stems than by leaves or twigs (Carter 1965, Fig. 30). Smaller-scale patterns of deposition were examined by Hirst & Stedman (1971), who used fluorescent particles in a wind-tunnel to show that most were caught on stems and spines, protruding veins and upwind edges of leaves. Other wind-tunnel experiments with needles of Sitka spruce, *Picea sitchensis,* showed that *Lycopodium* spores were more easily caught on the waxiest parts (Forster 1977). Older needles caught fewer spores, presumably because their wax was less rough.

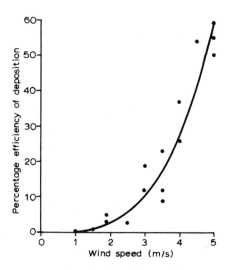

Fig. 30 − Variation with wind speed of the percentage efficiency of deposition on a square-section rod by ascospores of *Eutypa armeniacae.* (After Carter 1965).

Many kinds of windborne pollen grains are large enough (diameters 10-30 μm) to bump into the stigmas of flowering plants at low and moderate wind speeds. Such stigmas tend to be large, bare and well exposed above or away from leaf masses; they also tend to be sticky and branched, even feathery − all characteristics that increase the chance of catching windborne pollen. Some fungi also catch windborne pollen. *Retiarus* grows on the leaf surfaces of some evergreen trees and shrubs in South Africa. It sends up short, pointed branches that trap the grains, which are then fed upon, and spores produced later are spread in rain splash (Olivier 1978). Bacteria are perhaps too small to be caught by impaction, but they are known to be carried on pollen grains.

Vegetation can be an effective filter of windborne organisms. In winds stronger than a few kilometres an hour, impaction dominates fall-out so that the rate of flow of organisms to the ground becomes greater than that expected from fall-out alone. Because the rate of deposition increases with both gustiness and mean wind speed it varies between day and night – a maximum can often be expected in the afternoon, and a minimum around dawn. Rate of deposition also varies with the properties of the vegetation canopy, especially its depth, density and movement. Because both gustiness and mean wind speed decrease quickly downwards into a canopy, deposition rate is probably greatest in the canopy top. This has been demonstrated by Aylor (1975b) using stained pollen of ragweed, *Ambrosia elatior,* caught on maize, *Zea mays*. Inside a forest, there is often little wind, and impaction plays a small role in deposition compared with fall-out. Where the vegetation canopy bends in the wind, and where leaves are made to flutter, then the area able to catch windborne organisms will vary with wind speed. Even in light winds, leaf attitude may change with the direction of the sun and therefore perhaps also with the wind.

2.3 LANDING BY INSECTS

It seems likely that flying insects usually control their landing: they land when their mood for flight gives way to a mood for landing. A change of mood may come about after a long flight, or when the weather becomes too cool or windy, or when the insect seeks a goal, such as shelter, food, a mate or an egg-laying site. To avoid a crash, landing must take place within the insects' boundary layer. There will be occasions, particularly with small, slow insects in strong winds, when this can be found only within a vegetation canopy, or in sheltered places such as behind an obstacle or in a hollow, or in a gap in a crop. Where the wind blows into the edge of a crop, not only does the mean speed decrease quickly into the crop but also sheltered patches develop in the lee of individual plants. Windborne insects therefore tend to land both on the windward side of a crop, and behind individual plants. Accumulation on windward sides has been demonstrated for several kinds of insect. Taylor & Johnson (1954) found this to be so for colonies of the **bean aphid**, *Aphis fabae*, following spring migration from spindle trees, *Euonymus* (Fig. 31). Kieckhefer & Medler (1966) used sweep nets to catch **potato leafhopper**, *Empoasca fabae,* and found consistently larger catches on the edges compared with the middle of an alfalfa field. They also found that the leafhoppers were more abundant in the higher parts of a gently undulating field, and attributed that to difference in temperature at the time of egg laying (due to cold in the hollows); but such a distribution would also be expected if gravid females were differentially trapped in the higher, more exposed parts. These results suggest places that might be checked for first arrivals of windborne pests. Migrant **damson-hop aphids,** *Phorodon humuli,* were shown by Campbell (1977) to be more abundant on the leeward side of hop plants, *Humulus,* reflecting the pattern of deposition of probing migrants flying in spring from *Prunus* species.

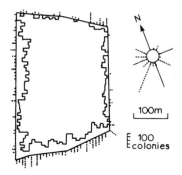

Fig. 31 – Distribution of bean aphids, *Aphis fabae,* around the edges of a bean field at Sutton Bonington, England, 1952. On the southern side there was a 1 m high hedge, but the other sides were well exposed. Edges were sampled in 5 m sections, 2 m into an autumn-sown crop, taking two stems at random in each section. Dots outside the edges show counts of primary spring migrants, arriving in late May and early June, mainly on winds from between south-west and north-west (frequencies of daytime winds during the invasion period shown by the wind rose). Bars inside the edges show the numbers of colonies 1.2 cm or more in length. The southern edge was the most heavily infested with both migrants and colonies. (After Taylor & Johnson 1954).

Near a hedge or other forms of **windbreak**, zones of shelter are produced, their size and intensity depending on porosity (Fig. 15d). Where wind speed is reduced to less than insect air speed, controlled landing is possible, leading to maximum accumulations at a distance downwind from the windbreak equal to about 2–4 times its height. But drifting particles, such as pieces of paper, gather in the same places as do **aphids, thrips** and **scale insects** (Lewis 1969a, 1973, Greathead 1972), so it is not clear to what extent such small insects actively control their landing, other than to fall out where they find themselves in light winds. Lewis (1965a) examined the landing of **lettuce root aphid,** *Pamphigus bursarius,* near a windbreak. This aphid sucks lettuce roots, thereby causing wilting and sometimes death. In Britain, winged adults fly during early summer from galls on poplar trees (*Populus*) to land on the lettuce plants and give birth to nymphs that burrow through the soil to the roots. *Winged* aphids are produced on lettuce only in the autumn, and since *wingless* aphids are unlikely to wander far in the soil, the pattern of wilting in a crop shows where the first winged aphids landed. A lettuce crop was sown on 9 June 1964 in rows about 0.6 m apart along the line and on both sides of an east-west windbreak formed of 1 m high, wooden, slatted fence with 45% open spaces. Seedlings sprouted between 17 and 23 June, and plants were fully grown by 14 August. Suction traps caught flying aphids 22–29 June in winds from between south-west and north-west, mostly on 24 June. Wilting started on 25 July, and each row was checked on 30 July, and on 5, 14 and 21 August. Wilting was greatest along a line 2–3 m to leeward of the windbreaks (Fig. 32), the line thus showing where a heightening of density of airborne aphids in June is likely to have occurred due to sheltering by the windbreak (see also page 38).

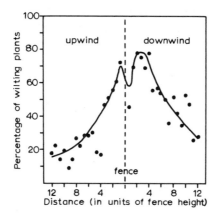

Fig. 32 – Percentage of wilting among rows of lettuce plants following a spring migration of the aphid *Pamphigus bursarius*, and its variation with distance upwind and downwind from a 45% slatted wooden fence – compare with Fig. 15d. (After Lewis 1965a).

Similar results were found by Lewis (1966) with **turnip mild yellows,** a disease due to a virus carried by the **green peach aphid,** *Myzus persicae,* Artificial windbreaks consisting of 45% slatted fences were set up facing south and west in an open field sown on 10 May 1965, and plants were examined for incidence of disease on 7 July. The vector had been first caught in a suction trap one day after sowing, so plants had probably been exposed to infection soon after germination. It was found that infected plants were most common behind the south-facing fence, at distances between one and four times the fence height. Winds blowing most directly on to the fences were from the south, consistent with accumulation of flying migrants on the leeward side.

The difficulty of landing at an exposed place when wind speed is greater than the insect's air speed is high-lighted by the marked drop of catch size in light and pheromone traps when winds become that strong (pages 102 and 150).

2.4 EFFECTS OF RAIN

Falling **rain drops** can sweep up airborne organisms in much the same way as vegetation filters them out – a mechanism known as **wash-out**, or scrubbing.

Collection efficiency varies with the sizes of both drop and organism. For a given size of organism, both large and small drops are poor collectors: the former because the airflow tends to sweep the organisms around; the latter because they fall slowly. Drops with maximum efficiency have an intermediate size, about 1 mm diameter, as has been shown by both laboratory observations and theoretical calculations (Star & Mason 1966). For drops of a given size, large organisms are more easily washed out because their inertia reduces the chance of being carried around the drops. Few airborne organisms with diameters less than about 5 μm are washed out by rain drops of any size; hence bacteria are unlikely to be washed out by rain (Graham et al. 1977). How such small organisms leave the atmosphere is still in doubt; perhaps coagulation is needed before they become large enough to be washed or filtered out. For somewhat larger organisms, with diameters up to a few tens of micrometres, wash-out probably dominates impaction as the mechanism by which they are brought to the ground, particularly for those that are being carried on winds hundreds or thousands of metres above the ground. Indeed, it should be remembered that rain can bring down organisms from aloft when there are none in the air near the ground. Starr (1967) has calculated the time needed for rain to wash out half the airborne organisms in a cloud — it is clearly greater for smaller organisms and smaller drops. In light rain (1 mm/h) of small drops, about 4 h is needed for 5 μm diameter organisms, and 1 h for those of 10 μm diameter; in heavy rain (5 mm/h) of larger drops, the times are 1 and 0.25 h. Because a windborne cloud of organisms may take as long as 10 h to pass through an area of long-lasting rain, its volume density will be greatly reduced. Calculations show that the relative importance of impaction over wash-out increases as cloud density decreases (Scriven & Fisher 1975); hence impaction can become the dominant landing mechanism before a cloud of organisms has passed through an area of long-lasting rain.

Although direct collision probably dominates in wash-out — not only on to the lower surface of a falling drop but also on to its upper surface due to the presence of a turbulent wake — other mechanisms may play a role. These may include the effects of electrostatic forces, either because the organisms themselves are charged or there is an atmospheric electric field, and of phoretic forces — for example, down the temperature and water vapour pressure gradients close to drops growing by condensation or shrinking by evaporation (Grover et al. 1977).

Rain drops striking the ground or vegetation often splash; hence some organisms that they have swept up will be made airborne again. The droplets carrying them are more easily windborne and they may evaporate. Even those organisms that stay on the ground or vegetation in a water film may become airborne again when later drops strike and splash the film. Wettable spores, those able to get inside a falling rain drop, may behave differently when splashed than do those that are non-wettable and remain on the drop's surface.

2.5 DEATH IN THE AIR

Although all windborne organisms sooner or later come back to the ground, some will have died before doing so. But éven dead organisms can be harmful – for example, by causing allergies. A windborne organism may be killed by some toxic part of the air, or by drying out, or by sudden wetting, or by sunshine, or even by being caught and eaten. Little is known about the causes of death of flying **insects**, but both coldness and dryness no doubt play their parts. In laboratory studies on the effects of varying temperature and relative humidity on the survival of crawlers of the **California red scale,** *Aonidiella aurantii,* by measuring their subsequent ability to feed and form scales on pieces of lemon leaf, Willard (1973) found that the mean survival period at 15°C was 17 h at 70% and 14 h at 35% relative humidity, whereas at 35°C these periods were more than halved. Such long periods leave little doubt that crawlers could withstand windborne movement for 100 km or more in wind speeds they are often likely to meet (page 90). With **bacteria**, small amounts of some atmospheric impurity (the 'open air factor') seem to be very effective in killing them. One source of such impurity seems to be a product of the chemical reaction between atmospheric ozone and olefins shed into town air, perhaps from petrol (Druett 1973). Bacteria are also sensitive to changes in temperature and relative humidity. This has been demonstrated in the laboratory with suspensions of *Serratia marcescens* and *Pasteurella pestis* atomised into air whose relative humidity was changed suddenly by mixing with drier or moister air (Hatch & Dimmick 1966). Similar experiments have been made with *Flavobacterium* showing that increased temperature is more of a killer than humidity changes, at least above 25% (Ehrlich *et al.* 1970, Fig. 33). But survival is no doubt enhanced by their being

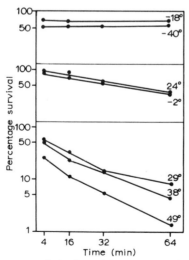

Fig. 33 – Percentage survival of an airborne cloud of *Flavobacterium,* and its variation with time at the temperatures shown. (After Ehrlich *et al.* 1970).

carried in rain splash droplets: both *Pseudomonas glycinae,* the cause of **bacterial blight of soybean,** *Glycine max* (Venette & Kennedy 1975), and *Erwinia amylovora,* the cause of **fireblight** in pears and apples (Schroth *et al.* 1974), are carried this way.

Virus particles are sensitive to drying out, possibly brought about by some change in the protein coating around the nucleic acid core. Donaldson (1972) found a rapid decline in infectivity of saliva droplets containing **foot-and-mouth virus** as relative humidity decreased. 'Open air factor' seems to have little effect on this virus (Donaldson & Ferris 1975), and the droplets seem to protect the virus from inactivation by sunlight.

Some viruses and fungal **spores** are sensitive to sunlight. For example, the ability of summer spores of **yellow rust,** *Puccinia striiformis,* to germinate is very greatly reduced after only one day's exposure (Maddison & Manners 1972, Rapilly 1979). Long-distance movement of this species is therefore more likely in cloudy and even rainy weather than in sunny and dry weather (see page 29). Other fungal spores are easily killed by dry air: conidia of **bluestain** fungus, *Ceratocystis,* rapidly lose viability in relative humidities less than 95% (Dowding 1969), whereas with *Calonectria crotalariae,* a fungus causing **black rot disease in peanuts,** *Arachis hypogaea,* less than 10% germinated at 73% and 33°C after 2 min exposure, and less than 0.1% after 30 min (Rowe & Beute 1975). Survival of *Mycoplasma gallisepticum,* the cause of **chronic respiratory disease in poultry,** also varies with relative humidity (Wright *et al.* 1968).

CHAPTER 3

Downwind drift

Once an organism is airborne it tends to drift downwind. If the organism has some means of moving *through* the air, rather than simply *with* it whilst falling, then its drift can be offset, wholly or in part. Of all the many kinds of small organisms that can become airborne, winged insects are the most able to move through the air — by flapping their wings. Insect flight is considered in Chapters 4 to 7; here we are concerned with drift.

The *distance* moved by an airborne organism is the product of the time spent in the air and the speed made good across the ground. Chapter 2 dealt with the *time* spent airborne; in this chapter we look at the displacements resulting from *movement* of the air.

3.1 WINDS AND TRAJECTORIES

The path traced out by an airborne organism is its **trajectory**. It is a more or less irregular line that can be either observable directly or estimated from the known or likely wind. Some idea of the kinds of paths traced out by downwind drift can be got by watching a zero-lift balloon. Such a balloon is inflated to stay at a fixed height in still air. In a wind, however, the balloon drifts away on a path made up of an endless series of short steps, each in a direction more or less different from the ones immediately before and after, and each with a different speed. This jerky movement is easily seen by looking downwind from the point of release, and it can be mapped photographically, using visual or radar devices. No two balloons follow exactly the same path, even when released only seconds apart. Soap bubbles, thistle down and blowing leaves behave in much the same way, and so also must many organisms that go unseen as they drift on the wind. It must be remembered, however, that airborne organisms differ from a zero-lift balloon in that they sink under the effect of gravity because they are denser than the air they displace. But sinking can be offset by the air rising.

Trajectories are jerky because they are affected by wind gusts. On the space and time scales of an individual gust, the trajectory is comparatively simple; it is the combined effects of a sequence of gusts that results in the tortuous trajectory. Again, within a larger-scale wind system, upon which the smaller ones are superimposed, the trajectory is comparatively simple, for distortions due to the many smaller-scale systems can often be looked upon as more or less random deviations that add to the uncertainty of estimating a particular trajectory. The deviations must be taken as random because little or nothing is known about them individually when they are too small to be described adequately by the network of field stations measuring the wind. Such stations are usually tens of kilometres apart, often much more, so trajectories can be calculated with acceptable precision often only from observable wind systems on scales of hundreds or thousands of kilometres. Such systems are depicted on weather maps, whose construction, availability and use therefore need to be described. The structure of medium-scale systems such as squalls (page 50) and fronts (page 52) can often be estimated from sparse field measurements and known or likely behaviour; but the smallest systems, such as gusts, are unresolvable, except by special observing networks.

3.1.1 Windfield maps

A windfield map depicts the wind over a given area at some given time. It can be used to estimate the wind (speed and direction) at any chosen place for the map time. A sequence of maps therefore enables an estimate to be made of the variation of wind with time at any chosen place, or a chain of places along any line, including a trajectory. Each map is based upon field observations from a worldwide network of stations, where winds are measured at the same fixed times each day. Some of these stations have been installed for the preparation of aviation weather forecasts, and observations from them are available within an hour or two worldwide by means of national and international communications provided for that purpose. Other stations have been installed for the preparation of climatological statistics as an aid to, for example, agriculture, forestry, hydrology and engineering, but individual observations are available, although usually only after some days, weeks or even months. Both kinds of stations usually provide wind measurements at some height varying from close to the ground up to ten metres (the internationally agreed height for aviation purposes). At only a few places in each country are there routine facilities for measuring winds at greater heights — by means of towers, tethered balloons or free-rising balloons. Winds can also be measured from cloud movement, as pictured by geostationary satellites, but the assumption must be made that clouds drift with the wind — an assumption that is not always true. Such satellite measurements are particularly useful in sparsely reported areas, such as oceans, deserts and tropical forests, but because of vertical shear the wind at cloud height may differ from that near the ground.

There are three ways by which an analysis of the windfield can be obtained: from the wind measurements themselves, from the atmospheric pressure field, and from both together. Each measurement is represented on the map by a 'wind arrow', using internationally agreed symbols (Fig. 34a). The arrow flies with the wind, with the number of 'feathers' indicating wind speed, and the arrow head lying at the place of observation. Alternatively, wind speed may be shown by a number. For example, at station A the wind is westerly (blowing *from* the west) with a speed of 5 metres a second (18 kilometres an hour), whereas at station B it is southerly at 12.5 metres a second. One large feather represents 5 metres a second, and a half feather is 2.5 metres a second. Once all available measurements have been plotted on the map, lines of constant speed, or **isotachs,** may be drawn by interpolation between the reporting stations, and the resulting pattern of lines shows areas with strong and weak winds. Lines of constant direction, or **isogons,** may be drawn similarly, and from them **streamlines** can be deduced. A streamline is everywhere parallel to the wind at places that it crosses. Fig. 34b shows an analysis of the measurements plotted on Fig. 34a. The isotachs suggest, by interpolation, that the strongest winds are about 17 metres a second, stronger than anything actually measured in this example because no observing station happened to be there. The streamlines show a progressive change of direction across the map. For examples of streamline maps, and the wind reports on which they are based, see Figs. 41, 74, 76, 82, 84, 88, 89 and 92.

Fig. 34 illustrates one of the sources of error in windfield analysis — the sparseness of observations. The denser the network, the greater the precision of analysis. Patches of strong or weak wind, or small waves in the streamlines can be easily overlooked if they are smaller than the network spacing. Possible errors in analysis that may come about through ignoring these unknown small wind systems must always be kept in mind when windfield maps are being used to estimate trajectories of windborne organisms. Fortunately, many such small wind systems are short-lived and are unlikely to lead to large errors in trajectories on scales of hundreds or thousands of kilometres. Fig. 34 also illustrates another source of error: unrepresentativeness of some individual observations. The aim of the map is to represent the windfield at some standard height (usually 10 m) *in open country*. This implies that measurements are made on short towers away from obstacles such as trees and buildings. Instruments are expensive, and large open sites in cities, forests and hilly country may be impossible to find. Hence some measurements are made at places that are not representative of the windfield at 10 m in open country. When streamlines and isotachs are drawn, allowance is made for possible unrepresentativeness (for example, station C in Fig. 34a), with smoothing introduced wherever it can be justified.

Another source of error is sampling time. Winds are almost always gusty (page 47), especially during the daytime. Fluctuations last from seconds to minutes, and when mean wind speeds are less than about 5 metres a second these fluctuations are relatively large, and direction can vary by $90°$ or more from the mean. In such weather, an instantaneous measurement, or even one averaged

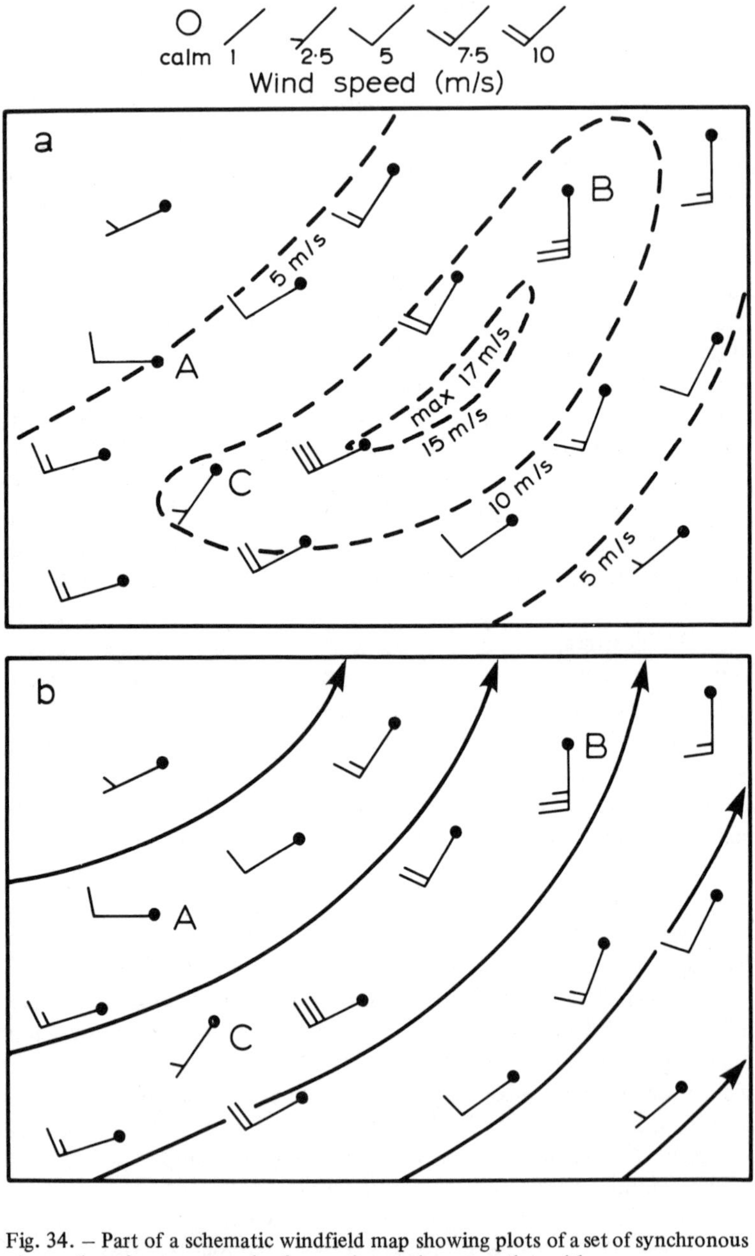

Fig. 34. – Part of a schematic windfield map showing plots of a set of synchronous observations from a network of reporting stations, together with:
    a   interpolated isotachs (lines of equal wind speed).
    b   interpolated streamlines (lines parallel to the wind direction).

over 10 or 20 seconds, can be unrepresentative of the large-scale windfield because it happens to be dominated by a transient, small-scale wind system.

Because there are these errors in measurement and analysis of winds, some smoothing is needed when drawing any set of streamlines or isotachs. It is possible to overcome them, however, at least in part, by analysing atmospheric pressure. Fields of wind and pressure corrected to some standard height are related (page 37), except near the equator or where they are changing rapidly. Broadly, the wind blows parallel to the **isobars** (with lower pressure on the left-hand side in the northern hemisphere), and it blows stronger where isobars are closer. Thus, a map of the pressure field, based on measurements at the network of stations already discussed, can be used to estimate both speed and direction of the wind at any place, bearing in mind the effects of ground friction (pages 38, 45; and see Fig. 80 for an example). The usefulness of a pressure field map for estimating the windfield depends upon the accuracy of the pressure measurements (including their correction to a standard height, usually sea level), their representativeness (especially in areas with large horizontal differences in air temperature), the effects of ground roughness, and the time changes in the isobar patterns. As regards the last, the spacing and curvature of isobars are always changing, and where the pressure patterns form and decay within hours (rather than days, as is usual with large-scale systems) the wind may blow strongly *across* the isobars.

Because the errors in estimating winds from a pressure field are not always easy to assess it is often worthwhile to combine wind and pressure analyses: the best compromise is then sought between the windfield based on measurements and the windfield estimated from the pressure field. In areas with sparse observations, greater emphasis is put on the likely fields within any given atmospheric disturbance, if it is of a kind that is better understood in other parts of the world with more plentiful observations.

Windfield analysis in detail is best undertaken by meteorologists who are aware of the uses to which the analysis will be put when preparing trajectories of windborne organisms. Nevertheless, windfield maps produced routinely for aviation weather forecasting are undoubtedly of great value in getting a first estimate of streamlines and trajectories. Maps covering a whole country can usually be provided every three hours, or even hourly, by most national meteorological services. Maps covering whole continents, or even a hemisphere, can usually be provided every 6 or 12 hours by the weather forecast offices at many international airfields. All such routine maps are filed, although for a more or less limited period, and can be consulted on request. Some meteorological services publish small-scale, simplified weather maps as daily weather reports, more or less in arrears, and they can be very useful for getting a first look at the large-scale wind systems present on a given day. Such published maps cover countries, continents or a hemisphere. Daily northern hemisphere maps, for example, are published by the meteorological services of USA, USSR, Japan and

West Germany. It should be remembered, however, that none of these maps has been prepared with estimation of trajectories in mind. It is therefore preferable to prepare maps from the original measurements, which are available from national meteorological services, and which can be supplemented by measurements from the already mentioned climatological stations, most of which are likely to be run by government departments (for example, agriculture, forestry, coast guard) or by universities.

3.1.2 Wind systems

Any windfield map, even for a small part of the earth's surface, is likely to show a seemingly bewildering array of streamlines or isobars, but closer examination will reveal a number of simpler, but overlapping disturbances on a range of scales. There is seen to be a great variety of **whirls, waves** and **windshift lines** (Fig. 35). Each whirl is more or less circular, and its streamlines show an instantaneous circulation of winds about a centre — turning either the same way as the earth, when it is called a **cyclone** (with an anticlockwise circulation in the northern hemisphere, clockwise in the southern); or turning the other way, when it is called an **anticyclone.** Where streamlines bend without forming a closed circulation there is a **wave,** which may also be looked upon as a weak whirl superimposed on a more uniform wind stream. Where two wind streams

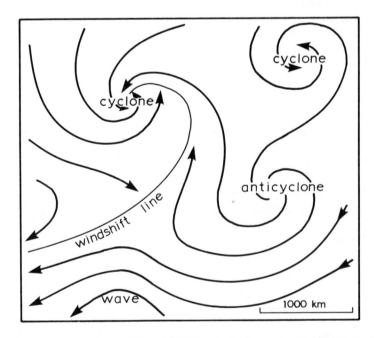

Fig. 35 — Part of a schematic windfield map showing streamlines to illustrate cyclonic and anticyclonic whirls, a wave, and a windshift line.

meet there can be a **windshift line,** across which there is a sudden change of direction, and often of speed. A series of windfield maps shows that individual systems form, grow, decline and fade away over periods of days and sometimes weeks, during which time they move across the earth's surface, sometimes splitting, sometimes joining with other systems. They vary in size, strength and speed. The overall picture is one of a continual churning of the atmosphere through an endless succession of moving large-scale eddies. The passing-by, build-up and decline of these eddies leads to the endless day-to-day changes in the wind at any given place. But these wind systems do not appear at random. A series of daily maps for many years will show that most form and die over particular parts of the earth's surface, and that the flow is more organised than might appear. Unlike the small and short-lived gusts that cannot be followed individually, it is possible to map each large-scale disturbance in more or less detail.

In middle latitudes, the winds are broadly westerly, whereas in tropical latitudes they are easterly (the **trade winds**), and between them is a belt of large, anticyclonic whirls (Fig. 36). The trade winds of opposite hemispheres meet near the equator at the **inter-tropical convergence zone** (ITCZ). Over land during summer, the ITCZ moves polewards and between it and the equator **monsoon** west winds develop. Within both middle-latitude westerlies and tropical easterlies, whirls, waves and windshift lines come and go. In northern hemisphere middle latitudes, cyclones are most commonly found moving north-eastwards in two belts across the oceans: one from south-eastern USA to northern Europe, the other from south-east of Japan to western Canada. West winds are strongest on their southern sides, and can take airborne organisms 1000 km or more in a

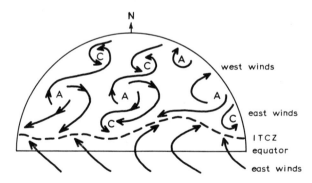

Fig. 36 – Schematic and simplified representation of the large-scale wind flow during the northern hemisphere summer, showing:
 east (trade) winds of low latitudes, with embedded cyclones and waves, and
    meeting at the inter-tropical convergence zone (ITCZ),
 west (monsoon) winds of low latitudes where the ITCZ lies well north of the
    equator,
 west winds of middle latitudes, with embedded cyclones and anticyclones.

day. Often they are accompanied by fronts (page 52) and windshift lines, and
sometimes there are moving anticyclones. There are spells of days, and even
weeks, when both cyclones and anticyclones move little, and winds are more
northerly or southerly than usual. Other slow-moving cyclones form over hot
lands bounded by cool seas, or over warm seas bounded by cool lands. Most
tropical cyclones and waves move westwards, the most vigorous being the
hurricanes of the Atlantic Ocean, and the typhoons of the Pacific Ocean.

Wind systems of many kinds are described in standard works on meteorology;
see, for example, Ramage (1971), Riley & Spolton (1974), Cole (1975), Donn
(1975) and Riehl (1979).

3.1.3 Downwind trajectories

The estimated trajectory of an airborne organism can be used to judge where
that organism has come from or might go to. It may be calculated by assuming,
at least in the first instance, that drift is downwind and at the wind speed. If
there is cross-wind flight on an assumed heading and at an assumed air speed,
another trajectory can be calculated, but if heading and air speed vary with time
in unknown ways then a trajectory cannot be calculated. Suppose Fig. 37a is
part of a windfield map showing two streamlines (continuous lines) and we wish
to estimate where an organism moving downwind at A has reached after a time
interval t. A first approximation to this trajectory is the line AB lying along the
interpolated streamline through A and of length Vt, where V is the wind speed
estimated from the isotach pattern. This would be a fair estimate if the streamline pattern remained unchanged during the interval t, but this would not generally be true. Instead, after time t, the streamline pattern may have changed to
that shown by the broken lines, and another approximation to the trajectory
would be the line AC, of length $V't$, where V' is the wind speed estimated from
the isotach pattern after time t. Another, and better, approximation would be

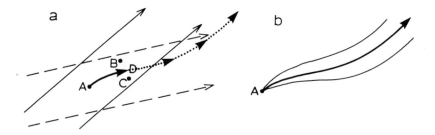

Fig. 37 – Construction of a downwind trajectory.
  a Segment, *AD*, derived from two successive windfield maps – initial stream-
     lines shown continuous, and final ones broken. See text for explanation.
  b Illustrating the difference between the most likely trajectory and the
     range within which the real trajectory is likely to lie with 90% probability.

AD, where *D* can be taken as midway between *B* and *C,* if the streamline pattern changed steadily with time. The beginning of this trajectory is tangential to the streamline through *A* on the first windfield map, and its ending is tangential to the streamline through *D* on the second windfield map. This graphical computation can be extended another step, by starting at point *D* and using a third windfield map for another time interval *t.* In this way a trajectory can be built up from a succession of segments. Ideally, the time interval should be as short as possible, but the shorter is the interval the more laborious is the computation. In practice, maps are seldom available more frequently than hourly. Moreover, the map need be changed only if the segment changes direction by more than, say, 20°. The greater is the change in segment direction, the less accurate is the calculated trajectory.

Trajectory computation can be automated by means of computers and numerical models of the windfield. Often, however, only an approximate trajectory is needed, and it can be estimated by eye from one map, or at most a few. If the large-scale wind systems are slow-changing over a day or two, trajectories longer than 1000 km can be estimated with an accuracy sufficient for a first look.

Because there is usually vertical wind shear within the atmospheric layer where the organisms might be drifting, it may be useful to calculate trajectories for several heights above the ground. Near mountains, coasts and rainstorms, where wind systems can be complex and quick-changing (pages 50–55), trajectories are difficult to calculate with accuracy. By comparing calculated and measured tracks of both constant-level balloons and chemical tracers it has been found that, in the absence of wind measurements at flight level, surface winds by day often give a good fit in middle latitudes, if they are doubled in speed and veered by 10°. (See Pack *et al.* (1978) for a discussion of trajectories over distances up to several hundred kilometres.)

Errors in calculating trajectories based on downwind movement arise mostly from two sources: uncertainties in windfield analysis (page 69) and in timing of the organism's movement. There will always be some doubt about the accuracy of a wind at a given place interpolated from a windfield analysis, for the analysis is based on only a sample of measurements, each of which may have some degree of error or unrepresentativeness. The accuracy of a trajectory will vary with place and time, but on many occasions there would be a 90% probability of the real trajectory lying within a fan-shaped area having an apical angle of about 20° (Fig. 37b). Note that the absolute magnitude of the error increases with distance along the trajectory. Errors are particularly large wherever there is a marked movement or distortion of streamlines from one windfield map to the next. In Fig. 37a, *BC* can then be large compared with both *AB* and *AC,* and there is greater doubt than usual where to place *D* along *BC.* Where such quickly turning winds are also weak (as in the centre of an anticyclone), any lack of precision in calculating the trajectory is offset by its shortness. By contrast, near a windshift line, where quickly-turning winds can also be strong, trajectories may appear to be discontinuous across the line. At other times, a better estimate can be got by shortening the time interval between maps. If maps are not available at sufficiently short intervals it may be necessary to estimate intermediate windfields by interpolation between known windfields. But where there are strong mountain, coast or rainstorm wind systems, errors can become unacceptably large in the absence of necessarily frequent measurements from a dense network of stations.

The start of a trajectory may not be known precisely. For example, the date an insect was seen or caught may not be the same as the date of arrival, for it may have been resting, feeding, egg-laying, or for other reasons have been flying little for a day or two. When many individuals are seen to arrive together and suddenly, however, the times of sighting and arrival can be taken to be the same. Again, only the approximate date of arrival of disease-carrying insects can be estimated from the timing of the subsequent disease. Even if the day is known, the hour of arrival may be critical because of variations in wind during the day. In such cases, it may be necessary to calculate several trajectories during the period of possible arrivals. If only one or a few of these trajectories pass over known or likely sources of the airborne organisms, then the trajectories can be used to get a better estimate of the arrival time. Where trajectories vary with height, because of vertical wind shear, light may be thrown on not only the most likely arrival time but also the most likely height at which movement took place.

All these trajectories assume *horizontal* winds. In fact, winds often have *upward* or *downward* components. Above the turbulent mixing layer near the ground, these components are almost always small — less than 10 cm a second, but their persistence for several hours can take slow-falling organisms up or down by a kilometre or more, through wind patterns that can vary markedly with height. Such three-dimensional trajectories, in the more or less smooth flow of the atmosphere above the surface turbulent layer, can be calculated by using windfield analyses on maps representing not *level* surfaces but **isentropic** surfaces. An isentropic surface is one which the **potential** temperature is everywhere the same. The potential temperature of a given mass of air is the temperature it would have if brought to a standard pressure (usually 1000 mbar), by compression or expansion, without adding or taking away any heat. For periods up to a day or two, air above the surface convective layer moves with little change of its potential temperature; hence winds there blow along isentropic surfaces. It is possible to calculate isentropic trajectories. For example, Christie & Ritchie (1969) examined the likely origins of a peak in pollen counts in May 1965 at Churchill, Manitoba, by using isentropic trajectories and showed that the pollen came more likely within the convective boundary layer than along isentropic surfaces at greater heights. On other occasions, downwind drift may well be in isentropic surfaces rather than in the horizontal.

3.2 EXAMPLES OF DOWNWIND DRIFT

Many kinds of organisms that are unable to keep themselves airborne have been found in the air: not only spores, pollen, viruses and bacteria, but also larger ones, such as nematodes, mites, spiders and wingless insects. Once airborne, each organism drifts downwind like a balloon. But whereas a balloon is usually less dense than the air it displaces (that is, it is buoyant), organisms are denser than air and will sink, unless the air is itself moving upwards fast enough. An airborne organism that is unable to generate enough lift to prevent its sinking can remain airborne only in the presence of sufficiently strong upcurrents, such as are associated with gustiness or the more persistent and widespread updraughts of larger wind systems.

We now consider some examples of downwind drift on various time and space scales and for various parts of the world.

3.2.1 Spores

The spores of a great many species of phytopathogenic fungi are known or thought to be carried on the wind, either alone, in clumps, or in droplets from rain splash or drip. They cause various spot, stripe, rust, smut, wilt, rot, blight and mildew diseases of crop plants. There are grounds for thinking that at least some kinds of spores can be carried thousands or even tens of thousands of kilometres on the wind. Thus, spores are found over oceans (for example, Erdtman 1937, Polunin 1955) and polar latitudes (for example, Pady & Kelly 1953, Polunin 1954, Ritchie & Lichti-Federovich 1967), far from their sources, on plants and in the soil. Spores are also found at heights greater than 3-6 km (for example, Meier & Artschwager 1938). Because about one day is needed for a spore to fall 1 km at 1 cm/s (a fall speed that is large for most spores, but small for most pollen grains), drift during fall would be about 1,000 km in a wind of, say, 10 m/s, a speed often to be found above the planetary boundary layer (page 38). For a slower fall speed, a spore starting from a height of 5 km could drift 10,000 km. Moreover, the kinds of spores reaching a given place seem to change with windfields on synoptic and global scales (for example, Kelly & Pady 1953). The following three examples show how the spread of fungal diseases can be on a *continental* scale.

Coffee leaf rust is caused by *Hemileia vastatrix*. It causes severe leaf drop, branch dieback and even tree death, and it has for long been widespread in Asia and Africa, causing serious loss of yield through defoliation. It led to the abandonment of coffee growing in Sri Lanka, and replacement by tea. The New World seemed to be free until January 1970, when the disease was discovered near Itabuna in Bahia state of Brazil. Surveys later in that year showed it was present throughout a wide but scattered coffee-growing area, and it may have been spreading for several years. Despite very extensive containment programmes, the disease later extended south-westward, reaching northern São Paulo state by January 1971, and Paraguay and northern Argentina by 1974 (Waller 1979;

see Fig. 38). Such a spread is in the direction of the dominant wind, and spores were in fact trapped by aircraft over the state of Paraná before the disease was reported there (Schieber 1975). By 1978 it had reached Peru, and Bolivia by 1980 (Waller 1981, Wellman & Echandi 1981). In Nicaragua, following discovery in Curazo district in 1976, disease spread was westward, again in the direction of the dominant wind. El Salvador was reached by 1979, Honduras by 1980,

Fig. 38 – Location map showing places mentioned in the description of the spread of coffee rust (page 77).

and Guatemala and southern Mexico by 1981 (Waller 1981). This evidence favours windborne spread of the spores, and it is supported by the observation of airborne spores in and above coffee trees in Kenya. It had been assumed that spread to Sri Lanka had been from East Africa on south-west monsoon winds, but Nutman & Roberts (1970) claimed that rain spread the disease, citing among other evidence the obvious relation between secondary and primary pustules due to rain water trickle along the undersides of leaves, and the spread of spores only after the first rain shower of at least 7.5 mm. Spores could become airborne, however, by evaporation of minute, infected splash droplets, if not by leaf flutter (page 23). Whether the disease reached Brazil on the wind from Africa is unknown, but there is some circumstantial evidence that it could have done so (Bowden *et al.* 1971):

- the disease was seen in Angola in 1966
- spores have fall speeds about 1 cm/s and in the presence of gustiness some could therefore be airborne for about a week if first carried up to a height of 1 km
- mean wind speeds in the lowest 3 km can carry spores across the Atlantic Ocean in about a week.

Maize rust, caused by *Puccinia polysora,* seems to belong to the New World, where American maize is little affected. In 1949 the fungus suddenly came to Sierra Leone, West Africa, almost certainly due to accidental introduction by man, and laid waste to maize there because it was more open to attack. The fungus then spread eastward, crossing the whole of West Africa by 1951, East Africa by 1952, and reaching Madagascar by 1953 (Rainey 1973; see Fig. 39).

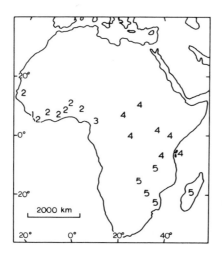

Fig. 39 – Spread of maize rust, *Puccinia polysora,* in Africa: from Sierra Leone in 1949 (year 1) to Mozambique in 1953 (year 5). (After Rainey 1973).

It seems to have spread on the wind during the maize-growing season — when monsoon south-westerlies were blowing over West Africa, and when north-easterlies were blowing over East Africa. By contrast, on a very much smaller scale, Mahindapala (1978) has demonstrated downwind spread to 25 m within a maize plot.

Two species of **poplar rust,** *Melampsora,* were first seen in New Zealand in March 1973. It is very unlikely to have come on cuttings because they are subject to strict quarantine regulations. Moreover, the pattern of subsequent spread through New Zealand suggests that the rust uredospores are readily windborne and infective far from their sources. First sighting at two places 450 km apart suggests an outside source. Both species had been present in Australia by February 1973, and had become widespread in New South Wales. From the size of the infection at first sighting, it seems that rust spores came to New Zealand in late February or early March. But winds then were mostly easterly over the Tasman Sea, and there were westerlies only from 1 to 3 March. A forward-tracking from New South Wales starting 1 March reached New Zealand after 2-3 days (Fig. 40), so it seems very likely that the rust spores travelled more than 3000 km over the sea (Wilkinson & Spiers 1976; see also Figs. 78 and 79).

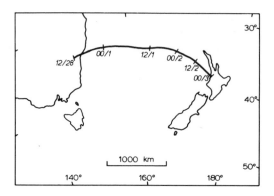

Fig. 40 – Trajectory of spores of poplar rust, *Melampsora,* from New South Wales to New Zealand, 28 February to 3 March 1973, assuming drift on winds at a height of 3,000 m. (After Wilkinson & Spiers 1976).

On the smaller scale of hundreds of kilometres, there have been some studies of spore drift over the North Sea. On 15 June 1962, south-west winds covered Britain and the North Sea. An aircraft fitted out to sample spores was flown over the southern North Sea. To obtain profiles of cloud density and spore type, but also to save the time needed for spiral climbs, a 'saw-tooth' plan was used with up-and-down flight on a fixed bearing. In this way the spore cloud was looked at up to a height of 1.8 km and along a downwind track of 350 km (Fig. 41). There were three kinds of spores: *Cladosporium,* pollens and 'damp-air' spores. (The first two take off mostly by day, and the last by night.) Each kind had lowest density about 100 km offshore, and greatest about 300 km offshore. Back-tracking gave the following most likely sources: at the north-eastern end of the flight – south-west England on the day before; at the south-western end – north-west France on the day before, or south-east England on the morning of 15 June. Densities tally with this and show that clouds of spores set free by day can be found far downwind. The cause of fewer spores being set free by night than might have been expected along the middle of the flight-track is unknown. Smaller densities near sea-level rather than aloft may have been due to settling on to the sea. Largest pollen densities were at a height of 500 m, and of *Clado-sporium* at 1,000 m, the lower height for pollens perhaps being caused by their falling faster than *Cladosporium* (Hirst *et al.* 1967).

Spread of spores across the North Sea has also been shown to be the cause of **powdery mildew of barley,** *Erysiphe graminis* f. sp. *hordei*, appearing in Denmark during the spring. Overwintering of conidia on winter barley is impossible there because the growing of winter barley has been prohibited, and field surveys

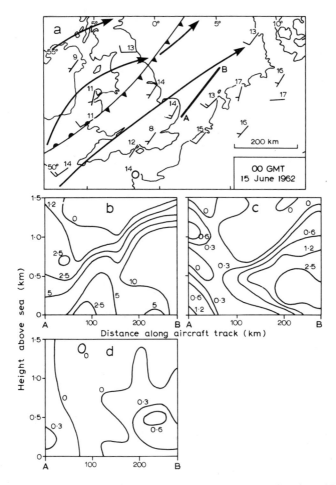

Fig. 41 – Vertical sections through spore clouds sampled by aircraft on 15 June 1962 over the North Sea along the track *AB* shown in a, which also shows geostrophic streamlines at 00 GMT. Units are spores/m^3. (After Hirst *et al.* 1967).
b *Cladosporium*   c pollens   d damp-air spores

found only rare overwintering on volunteer plants. To test the possibility of spores arriving from outside the country, they were trapped on the island of Sylt (55°N 8°E) in the spring of 1971 and 1972, and back-tracks were calculated using wind reports from a height of about 1.5 km (Hermansen & Stix 1975). With few exceptions, back-tracks that had crossed the British Isles and the nearby continental Europe had measurable numbers of conidia, and absence was attributed to rain over the presumed sources (conidia become airborne mainly during the day in dry weather). Further evidence was provided by experiments in 1972 and 1973 in which young barley plants, grown in isolated pots, were exposed for a few hours at three separate places in Denmark and then kept to see if infection developed (Hermansen *et al.* 1976). Again back-tracks on occasions when infective spores arrived were found to pass over likely sources where winter barley was being grown. This same species has also been shown to arrive as far north as 62°N, in the Faroe Islands, where barley plants were found to have become infected in late summer, the time of year when spore production is greatest in north-west Europe (Hermansen & Wiberg 1972).

Nagarajan *et al.* (1980) showed that the uredospores of **wheat leaf rust** in India, caused by *Puccinia recondita tritici,* oversummer in the Himalaya, and the appearance of the disease in midwinter along the foothills they attributed to spread on northerly katabatic winds blowing from the mountains (see page 54). Subsequent spread to north-western India is supposed to take place on southeast winds ahead of eastward-moving waves in spring, and spores are washed out by the accompanying rains.

Spores of the **white pine blister rust,** *Cronartium ribicola,* widespread in Europe and North America, are set free from *Ribes* (currant) plants at night and carried on the wind to white pines, *Pinus strobus,* and other five-needle pines. In northern Wisconsin, *Ribes* grows mostly on the edges of swampy hollows whereas white pines grow on low, sandy ridges between. There are grounds for thinking that the displacement of spores there is on weak night-time winds over a fetch of some hundreds of metres:

- coloured smoke showed that air drained downhill beneath the tree crowns, rose over the warm swamp water and then flowed back to the ridges;
- the spread of rust has a clear pattern: there is some on the tops of the tallest trees near the swamp, on the lower crowns further away, and low down on trees growing on the ridges.

Near lakes, weak night-time land breezes (page 54) seem to carry spores over the water, but the wind aloft brings them back and deposits them in downward flow 10–15 km inland. Such a displacement would throw light on the lack of rust near the shore, as well as much disease 10–15 km inland, because spore-laden air would flow *below* and *above* crowns near the shore. but *through* crowns of trees on the ridges (van Arsdel 1965).

Mention has been made of the possible spread of **coffee leaf rust** spores in rain splash droplets (page 78). Such a spread, to at least 77 m, was demonstrated with the fungus *Cytospora leucostoma* by setting up a series of traps in a line from a source in infected plum trees, *Prunus domestica* (Bertrand & English 1976). Numbers caught increased with wind speed during rainfall (Fig. 42).

Although there is considerable and convincing evidence for the reality of windborne spore movements over great distances, we might ask: why do fungal diseases not spread faster, further and more often than is observed? Spores are produced, and become airborne, in abundance, and suitable wind patterns are

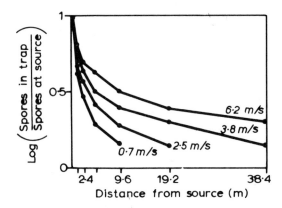

Fig. 42 – Variation of trap catch downwind from a source of *Cytospora leuco-stoma* at different wind speeds during rainy weather. Spores at source were recovered from run-off water. (After Bertrand & English 1976).

frequent, so why do the great majority of spores seem seldom to travel far? The answer is almost certainly that spores are, in fact, carried great distances frequently, but they seldom reach places where they can thrive and induce disease, even if they survive the windborne transport. Introduction of a new but susceptible host can increase the chances of disease development, giving the false impression of a sudden and unusual spread of spores, and the same effect can be produced by the emergence at source of a new and more virulent strain of fungus.

Studies of **pollen** drift supplement those of fungal spores. Pollen arrival in Shetland is an example. Shetland lies at $60°N$ $1°W$ with the nearest land being 250 km to the south-west and 400 km to the east. It is mostly moorland and there are almost no trees. Pollen caught in traps comes from rushes, sedges and grasses (in spring) and from heather (in summer); tree pollen can come from only far-away sources. During 1970, when 25 m^3 of air were sampled each day, tree pollen came in two main spells: 5-19 May (mostly birch, *Betula*) and 2-10 June (mostly pine, *Pinus*). Daily back-tracks for the latter spell ending at 00 GMT, based on six-hourly surface weather maps, showed that the peak catch on 9 June fitted with a short, swift displacement on east winds from Scandinavia (Fig. 43), whereas the much smaller catches on other days came with longer paths over the sea. There was no tree pollen on 11 June, which tallied with a sudden change to northerly winds behind a cold front (Tyldesley 1973).

Earlier observations from over the Atlantic Ocean nicely illustrate the variations in pollen composition according to source and wind systems (Erdtman 1937). From 29 May to 7 June 1927, pollen in the air over the North Atlantic Ocean was sampled daily during a westward passage of the M.S. *Drottningholm*.

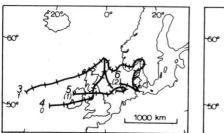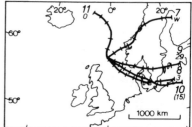

Fig. 43 – Day-to-day variation of trajectories to Shetland of pollen arriving
at 00 GMT from 3 to 11 June 1970 (dates shown at ends of trajectories). 6-h
segments except the last, which was 3-h. Average pollen concentration (grains/m³)
for 24 h periods (centred on arrival time) shown below each date. Estimated
numbers in brackets; *w* indicates no record due to wetting of spore trap in rain.
(After Tyldesley 1973).

Vacuum cleaners were used to draw air through filter papers at about one cubic
metre per minute. There were three peak catches (Fig. 44):

(a) sample I, when southerly winds were blowing across the North Sea, and
    most pollen was *Pinus* (most likely from a source over north-west Europe);
(b) sample V, when north-westerlies were blowing from Newfoundland behind
    a cold front that had passed the ship late on 3 June, and most pollen was of
    Alnus and Cyperaceae;

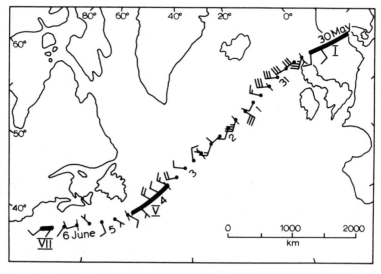

Fig. 44 – Route of the MS *Drottningham*, 30 May to 6 June 1927, showing wind
observations every 6 h, and locations of spore samples I, V and VII described in
the text. (After Erdtman 1937).

(c) sample VII, when south-westerlies were blowing offshore from north-eastern USA, and most pollen was of grasses (Graminae), *Plantago* and *Rumex*.

Elsewhere, cloud densities were smaller, but in a sample taken between Ireland and Iceland there was a noteworthy mingling of pollens from trees, shrubs and herbs. North-west winds were blowing around a cyclone moving slowly east from Iceland to Scandinavia, most likely bringing European pollen that had been airborne for at least a few days on a long curved trajectory over the open ocean.

An example of spread of pollen on south winds from North Africa to the south of France is given by Cour *et al.* (1980), who record the appearance on 19 May 1977 of pollen from several Saharan species, including *Argania, Calligonum, Ephedra, Fogania* and *Moltkia*.

Evidence for long-distance windborne movement of pollen comes from deposits in soil or mountain snow, or on vegetation. Moar (1969a) gives an example of *Casuarina* pollen found in recently fallen snow at an altitude of about 2000 m in the South Island of New Zealand. This tree is not native to New Zealand, and few are grown in gardens; hence the pollen is likely to have come from elsewhere. There was much red dust accompanying the grains, consistent with a source in Australia, where *Casuarina* is native. Moar (1969b) gives other examples from Antipodes Island (50°S 179°E) and Chatham Island (44°S 176°W), where pollen of the tree *Nothofagus fusca* has been found on vegetation, yet the nearest trees are in New Zealand, 800 km away. On a smaller scale, Tampieri *et al.* (1977) have demonstrated the windborne spread of pollen from chestnut, *Castanea sativa*, at least 40 km across the Po valley, northern Italy. On the other hand, in sheltered habitats, pollen spread on the wind can be very limited, with consequences in studies of genetic isolation. For example, in experiments by Handel (1976) with a sedge, *Carex platyphylla*, in the light winds of a New York State woodland, where staminate spikelets were marked with samarium before the anthers appeared (and the day after their appearance pistillate spikelets were collected, bombarded with neutrons and the radioactive decay of resulting europium measured), it was found that few marked pollen grains reached 1 m from source, and some flowers had none even at 50 cm.

3.2.2 Bacteria and viruses

Fireblight is a disease of apples and pears caused by the bacterium *Erwinia amylovora* (see page 31). Disease has for long been noticed to spread in the direction of the prevailing wind. Observations by Bauske (1967) in newly-infected trees of a block of seedling pears showed that both severity of infection and direction of spread seemed to be related to the degree of exposure to the prevailing wind. By shielding experimentally inoculated trees from the wind, the spread of disease was greatly reduced even though insects had access, suggesting that the wind can play a strong role. Experiments in a greenhouse with simulated wind and rain showed that bacteria could be carried at least 1 m from pears, and from *Pyracantha* and *Cotoneaster* (Bauske 1971). Mist-spraying of dwarf apples (to delay fruit bud and flower development by evaporational cooling until the risk of damage by frost has passed) enhances the spread of fireblight, presumably due to droplet impact (Spotts *et al.* 1976). The disease was widespread in Kent during 1969. In three orchards, the pattern of infection showed northwest to south-east lanes with diseased hawthorn, *Crataegus,* at the north-west ends (Glasscock 1971; see Fig. 45). Disease development was at the same stage in all trees, and infection therefore presumably took place at the same time, but the occasion is unknown. The relative roles of insects and windborne contaminated splash droplets in spreading fireblight is still not clear (Beer 1979, Billing 1980). Downwind spread of **coliform bacteria** in splash droplets has been demonstrated directly by setting up traps as far as 1,200 m from municipal sewage treatment, trickle-filter plants in the USA (Adams & Spendlove 1970). Similar experiments by Ragor & MacKay (1975) showed that coliform bacteria did not survive beyond about 20 m.

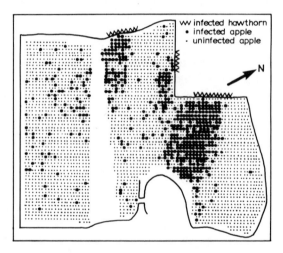

Fig. 45 – Apple orchard near Teynham, England showing distribution of fire-blight-infected trees in relation to sources on hawthorn hedges, 24 July 1969. (After Glasscock 1971).

Soil-living bacteria also become airborne. *Xanthomonas phaseoli* causes bacterial **leaf spot on bean,** *Phaseolus vulgaris* and *X. alfalfae* causes **leaf spot on alfalfa,** *Medicago sativa.* Wind-tunnel experiments with healthy plants and windblown natural soil or infected sand showed that disease incidence increased with wind speed and exposure time, and most lesions were on the lowest 10 cm of stem. In Kansas, both diseases are mostly on the south sides of stems, especially in hot and dry weather, and summer winds are mostly from the south (Claflin *et al.* 1973). Streptomycete bacteria in soil were found to become airborne, particularly on dusty days (for example, during cultivation), at the Waite Agricultural Research Institute, South Australia (Lloyd 1969).

Among windborne **virus** diseases, the one that has been most studied is **foot-and-mouth**. The virus causes disease in cattle, pigs and sheep. It has long been thought that the disease is spread by transporting sick animals or their milk, meat or wastes, and by such carriers as people and vehicles. Even so, there are strong grounds for thinking that spread can also be on the wind. Thus, between October 1967 and January 1968 there was a very worrying outbreak in Great Britain — animals on over 2,300 farms had the disease (Smith 1970, Hugh-Jones & Wright 1970). The outbreak started 8 km south-west of Oswestry (near 52°50′N 3°05′W), where sick pigs were found from 21 October onwards and were slaughtered on 26 October. The disease probably spread from this farm around 25-26 October, when it would have been at its most contagious. The majority of the outbreaks from 27 October to 1 November are likely to have come from this farm. Furthermore, most of these outbreaks were in the north-east, whilst winds between 22 and 26 October were from the south-west (Fig 46a). Three strange things about these outbreaks suggest that standing **gravity waves** in the atmosphere (page 41) helped the spread:

- the outbreaks were in three clusters;
- the clusters were about 20 km apart;
- outbreaks occurred within a week of the first (about the incubation period).

Waves were most probable on the afternoon and evening of 25 October. If they did help the spread this would explain the grouping and the long displacement downwind. Indeed, the speed of spread may have been due to the chance occurrence of standing gravity waves over the first outbreak (Tinline, 1970; see Fig. 46b). Viruses are set free in specks, mainly from the breathing tubes, and can drift on the wind for several hours before being breathed in by another animal. Spread coincides with cloudy, rainy weather which can be attributed to the better chances of viruses staying alive in moist air, and perhaps to wash-out by falling raindrops (page 63). Wash-out is more likely to help spread the disease afar because rain would sweep up the more easily displaced small specks rather than the larger ones, which would fall out on their own nearer the source. Once the moving of stock, people and vehicles has been banned following an outbreak, and the incubation period has passed, it is possible to obtain some idea of windborne spread, which will help in the search for new outbreaks. What is needed then is a knowledge of virus output rate (number and placing of sick animals), the windfield and humidity up to some tens of kilometres from known outbreaks, and the number and placing of healthy livestock (Sellers *et al.,* 1973; Sellers & Forman 1973). During outbreaks in Switzerland in 1966, the spread from valley to valley, and appearance on intervening ridges, supported windborne movement, rather than by vehicles or wild animals (Primault 1974). The next outbreak in Britain, during March 1981, seems to have been due to wind-

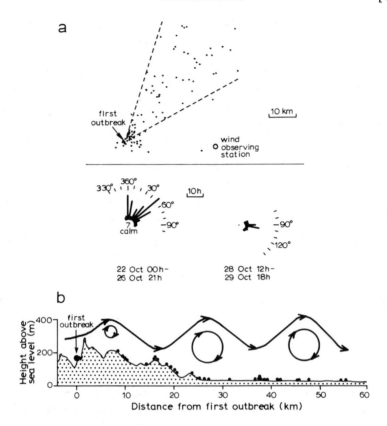

Fig. 46 – Foot-and-mouth disease near Oswestry, Wales, 1967.
   a Distribution of outbreaks 21 October to 1 November in relation to the
      first outbreak, and wind roses of hourly winds at the station shown. (After
      Hugh-Jones & Wright 1970).
   b Cross-section north-westward from the first outbreak, showing outbreaks
      25 October to 3 November in relation to lee waves likely on theoretical
      grounds to have been present on the afternoon and evening of 25 October.
      (After Tinline 1970).

borne spread of virus from outbreaks in Brittany, in north-western France,
judged by the similarity of virus strains (King *et al.* 1981).

    Fowl pest is another virus disease for which there is circumstantial evidence
for downwind spread (Smith 1964, 1970).

3.2.3 Mites and scale insects

Some agricultural pest species of tetranychid and eriophyid **mites** are carried by
the wind. These are tiny animals, often less than 1 mm across. Pady (1955)
caught *Eriophyes tulipae,* the vector of **wheat streak mosaic virus,** on spore
slides in Kansas, over 2 km from the nearest wheat, and especially when the

crop was being harvested. Greenhouse experiments (Slykhuis 1955) with a fan blowing across wheat seedlings, either infested with these mites or infected with the virus, showed that disease did not spread in the absence of mites and, although they were carried on the wind, disease spread only when mites were present on infected plants. This species also carries virus causing **kernel red streak of maize.** Nault & Styer (1969) set up sticky traps and caught many windborne mites in the autumn when the ears were drying out. *E. tulipae* can survive only on living host tissue and it must therefore move when its host dries out. Aphids can also carry away these mites (Gibson & Painter 1957). Another eriophyid mite, *Rhynacus breitlowi*, lives on the undersides of leaves of the evergreen *Magnolia grandiflora.* Laboratory experiments with a fan blowing through infected branches carried mites downwind when the wind speed was greater than 11 km/h (Davis 1964).

Among tetranychid species, the **citrus red mite,** *Panonchees citri*, was seen to reinvade a grove in California that had been cleared by spraying, by drifting on silk threads (Fleschner *et al.* 1956). The **avocado brown mite,** *Oligonychus punicae,* and the **6-spotted mite,** *Eotetranychus sexmaculatus,* were similarly seen to start leaving an avocado grove on evenings with very light winds. The **European red mite,** *Panonychus ulmi,* a major pest of deciduous fruit trees, and the **legume mite,** *Petrobia apicalis,* also drift on silk threads. Sticky traps downwind of a grapefruit grove in Texas caught **Texas citrus mites,** *Eutetranychus banksii,* at least 50 m away (Hoelscher 1967). Experiments with sticky traps in a bean field showed that the **2-spotted spider mite,** *Tetranychus telarius,* spread at least 1 m downwind, although this species does not use silk threads (Boyle 1957).

The predatory mite *Amblyseius fallacis,* which feeds on other pest mites in apple orchards, also drifts on the wind. Although it overwinters in the ground cover, it crawls up the tree trunks in spring and early summer, but later it also drifts, as shown by sticky traps, not only between trees but also to at least 70 m downwind (Johnson & Croft 1981).

All these experiments used sticky traps, which can be reached only by drift. But trap plants can also be used. For example, Thresh (1966) demonstrated movement particularly in the downwind direction by the **gall mite** *Phytoptus ribis,* the vector of **reversion** virus in blackcurrant, *Ribes.* Hexagonal rows of bushes had a source of reverted and infested bushes at their centre. Spread of disease was monitored by regular examination of the leaves, and at leaf-fall the positions and numbers of galls indicated the spread of mites (Fig. 47).

Several **scale insect** species are carried downwind. These are small, sap-feeding insects, the females of which are largely immobile. They are similar in size to mites, and the flattened bodies of the first instars (crawlers), with their long, fine hairs, reduce their fall speed to the order of 10 cm/s as they drift on the wind. The **pine tortoise scale,** *Toumeyella numismaticum,* is a soft scale of northern coniferous forests of North America, attacking and sometimes killing

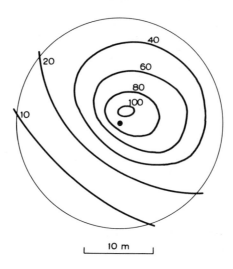

Fig. 47 – Distribution of galls at leaf fall on blackcurrent (mean number of galls/bush) caused by the mites *Phytoptus ribis* spreading on the wind from a source at the centre of an array of bushes. (After Thresh 1966).

young pine trees. Observations in Manitoba, using sticky traps, showed that they could be windborne up to at least 5 km from a severe infestation (Rabkin & Lejeune 1954). Similarly, in Alberta, the **pine needle scale**, *Phenacaspis pinifoliae*, was found to be carried at least 3 km (Brown 1958). The **sugarcane scale**, *Aulacaspis tegalensis* is a pest in parts of East Africa and south-east Asia. In experiments in Tanzania a line of sticky traps was set up over ploughed land up to one kilometre downwind of an area of infested sugar-cane. At one kilometre the cloud density was found to be $10^{-4}/m^3$, giving horizontal flows of scales more than enough to let infestations spread between fields, or even over much greater distances (Greathead 1972; see Fig. 48a).

The **California red scale**, *Aonidiella aurantii*, is a serious citrus pest around the world. Experiments with sticky traps at an abandoned and highly infested lemon grove in South Australia showed that crawlers could be carried at least 300 m (Willard 1974, 1976). The **beech scale**, *Cryptococcus fagisuga*, is also carried downwind; it allows the entry of a fungus, *Nectria*, that causes beech bark disease (Wainhouse 1979, 1980). The **elongate hemlock scale**, *Fiorinia externa*, has become a serious pest of eastern hemlock, *Tsuga canadensis*, in eastern North America. Again by using sticky traps, McClure (1977) showed that crawlers were taken at least 100 m downwind from an infested, isolated clump of trees in Connecticut. Eggs, gravid females, and needles infested with all stages can also be windborne (McClure 1979; see Fig. 48b). Similar experiments with the **red pine scale**, *Matsucoccus resinosae*, showed windborne movement at least 1.6 km (Stephens & Aylor 1978). Evidence has been presented by Hill (1980)

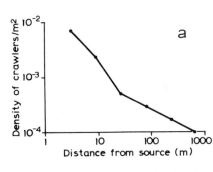 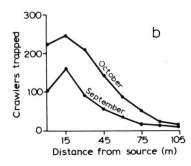

Fig. 48 – Variation of aerial density of scale crawlers with distance from source.
a Sugar-cane scale, *Aulacaspis tegalensis,* at the Arusha Chini Estate, Tanzania, numbers of crawlers blowing through unit cross-wind area in a vertical plane at a height of 1.5 m above ground in 3.5 h periods. (After Greathead 1972).
b Elongate hemlock scale, *Fiorinia externa,* downwind of a stand of infested trees in Connecticut; 14-day periods of trapping on 125 cm² surfaces at each distance, and means of four transects to the north-east, south-east, south-west and north-west. (After McClure 1977).

that crawlers of the **coccid** *Icerya seychellarum* on Aldabra atoll (9°S 46°E) are carried over water at least a few kilometres.

There is no doubt that all these mites and insects were carried on the wind; in no other way could they have reached the sticky traps.

Crawlers of the **coconut scale,** *Aspidiotus destructor,* are also carried on the wind, and so are those of the **mealybug,** *Pseudococcus njalensis.* This species is the main carrier of a virus causing **swollen shoot disease of cocoa** in western and central Africa. Its airborne displacement was studied by clearing a 30 m square plot of cocoa trees at Tafo, Ghana (6°15′N 0°20′W) and setting out a grid of seedlings to act as traps (Cornwell 1960). The fraction of seedlings becoming infested decreased downwind to at least 30 m from the edge of a stand of infested trees, thereby showing the possibility of a spread among trees other than by crawling through the touching crowns.

Even wingless forms of aphids can be windborne. For example, there was a notable early catch on 7 May 1966 at Brookings, South Dakota, of the **corn leaf aphid,** *Rhopalosiphum maidis,* and of the **apple-grain aphid,** *R. fitchii,* following a day of warm, south-east winds, an arrival resembling those of many species of flying insects discussed in Chapter 5 (Medler & Ghosh 1968).

3.2.4 Caterpillars

Of the larger windborne, wingless insects, the tree-infesting caterpillars of certain moths have been most studied. The **Douglas-fir tussock moth,** *Orgyia pseudotsugata,* is a defoliator of pines in North America. Epidemics occur about every 7-10 years almost simultaneously over its whole range, last about 3 years, and cause death, top kill or stunted growth. Female moths have tiny wings and cannot fly. After hatching, the first-instar caterpillars drop on silk threads up to 3 m long, which break and let the caterpillars drift away (see also page 58 and Fig. 28). Severe infestation is unlikely, however, beyond a few hundred metres (Beckwith 1978, Mason 1974), and caterpillars seem likely to reach that distance only if they are first carried above the canopy or beyond a forest edge (Williams *et al.* 1979). If caterpillars are to travel long distances in great numbers it seems that they must take to the air several times (Edmonds 1980). A simple model of population growth gives results consistent with observations although it does not involve migration (Berryman 1978). A similar species is the **Bruce spanworm,** *Operophtera bruceata,* a pest of many broad-leaved trees, especially in middle latitudes of North America. Again the female is flightless, and first-instar cater-pillars spin down on long silks. By means of sticky traps in lines from a stand of aspen, *Populus tremuloides,* Brown (1962) showed that the caterpillars could be caught to at least 750 m. Caterpillars of the **gypsy moth,** *Lymantria dispar,* are probably carried similar distances (Leonard 1971, see also page 19), but those of the **cotton leafworm,** *Spodoptera littoralis,* are probably carried less far, after hatching from eggs laid on objects (not host plants) high above the ground. Some **spider** species are also well known to become airborne. Field observations with sticky traps on the Netherlands coast (Wingerden & Vugt 1980) showed that *Erigone arctica* could be drifted on to the beach, where they died in a few days. The **plant bug** *Nysius groenlandicus* uses windborne seeds as a vector. Bocher (1975) found in Greenland that eggs are laid on the ripe fruits of *Dryas integrifolia,* the only plant with such seeds.

Lastly, we take note of soil-dwelling **nematode** worms. In strong, dry winds, when the soil is blowing, these microscopic nematodes became airborne. For example, *Trichodorus,* vectors of **tobacco rattle virus,** causing **spraing disease of potatoes** in Britain, have been found downwind of an infected potato field in Fife, Scotland (Cooper & Harrison 1973). Many species were recovered from drifted soil in Texas (Orr & Newton 1971).

CHAPTER 4

Insect flight within the boundary layer

An insect's flapping flight enables it to move *through* the air. Because the air is itself moving, the insect's displacement after a given time will be the sum of the displacements due to flight and to wind. In Fig. 49, the displacement due to flight is represented by the arrow *AB,* whose length is proportional to the **air speed** (the speed at which the insect would move in the absence of wind), and whose direction is the insect's **heading** (the direction in which it would move in the absence of a wind). On the same diagram, the displacement of the air during the same time interval is represented by the arrow *BC,* whose length is proportional to the **wind speed,** and whose direction is the **wind direction.** The

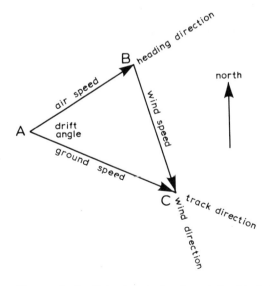

Fig. 49 – Movement of a flying insect, showing relationship between heading, track and wind.

insect's displacement across the ground is represented by the arrow *AC*, whose length is proportional to **ground speed,** and whose direction is the insect's **track** across the ground. Consider now two simple cases. First, if there is no wind (*BC* in Fig. 49 vanishes), *AB* and *AC* coincide, and the insect's displacement is determined solely by its ability to move through the air. Second, if the insect has no ability to fly (*AB* vanishes), *BC* and *AC* coincide, and the insect's displacement is determined solely by the wind (downwind drift). There is always some wind, and heading can be in any direction relative to that of the wind. As a result, *AB* and *AC* have different directions, and their difference θ, is the **drift angle.**

Direct observation of insect trajectories by eye (assisted, perhaps, by magnifying or night-viewing devices) is always limited: even large insects can usually be traced across country at most only some hundreds of metres. Radar can extend the range. For a given radar set, range increases with insect size. Entomological radars used by Schaefer and by Riley can detect individual locusts at a range of a few kilometres, and 1 cm-long moths at about 1 km (Schaefer 1976, Reid *et al.* 1979, Greenbank, *et al.* 1980, Riley *et al.* 1981; and see pages 126, 163). *Clouds* of insects are, of course, more conspicuous, and can be followed much further than individuals, both by eye and by radar.

When wind speed is less than air speed, an insect can head into wind and make progress across the ground — that is, it can reach a goal, even though compensation for wind may have to be made; heading and track are in general different (Fig. 49). Such a goal may be, for example, food, shelter, a mate, or an egg-laying site. When wind speed is greater than air speed, no progress can be made against the wind — the insect would be blown backwards. Hence an upwind goal cannot be reached. An insect finding itself in such a strong wind will usually settle, seek a sheltered spot with a lighter wind, or turn and head downwind.

The air speeds of most insect species are unknown, but some have been measured in the laboratory, and a few in the field. By and large, air speed is greater for larger insects (Fig. 50), varying up to about 10 m/s (35 km/h).

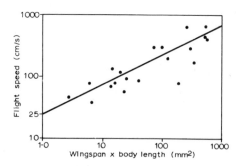

Fig. 50 – Variation of air speed with insect size (wing span X body length). (After Lewis & Taylor 1967).

Heading and ground speed are seldom noted in the field, although tracks may be. Wind speed and direction are also seldom noted, in the mistaken belief that a measurement somewhere nearby can be used in place of one where the insect is flying. Radar is a powerful tool for measuring track and ground speed (through multiple exposure photography), and air speed and heading can then be calculated if the wind is also known (Schaefer 1976).

Because the wind speed often strengthens upwards through a layer at least some hundreds of metres deep (the **planetary boundary layer** – see page 38), for a given insect, place and time there may be a height above which wind speed is greater than air speed. Below that height there is a layer in which the insect could reach any goal; this layer is known as the **insect boundary layer** (Taylor 1958, 1960, 1974), it should not be confused with the planetary boundary layer. Flight within the insect boundary layer often differs greatly from that above the layer, even for the same individual. Flight above the insect boundary layer is considered in Chapter 5.

Insect flight is almost always watched close to the ground, and is therefore often likely to be within the boundary layer. Many insect species may well have more or less clear-cut times in their life when flight is either mostly within or mostly above the boundary layer. Thus, the so-called "appetitive' flights, goal-seeking, are likely to be within the boundary layer because an individual can then fly to any point on or near the ground, wherever it may be, and if it is within flight duration. Some species, such as butterflies, travel long distances in more or less fixed compass directions, with compensation for changes in wind; others, such as some flies, make seemingly random, fast, searching flights, with frequent changes in direction; yet others, such as bees, return to known goals; and others fly upwind in an odour plume. We consider now some examples of these kinds of flight.

4.1 PERSISTENT CROSS-COUNTRY FLIGHT

The **monarch butterfly**, *Danaus plexippus*, occurs widely over the USA and southern Canada. Its caterpillars feed on a variety of species of milkweed, *Asclepias*. It migrates southward in autumn to overwinter along the Gulf of Mexico coast, the California coast, and in central America (Urquhart 1960, Urquhart & Urquhart 1977), the same individuals returning northward rapidly during the following spring. For many years, butterflies have been tagged with code numbers to trace their movements. It was found that those of the eastern population, from around the Great Lakes, seem to move south to the Gulf coast and then west to the Sierra Madre Occidentale in Mexico, where they overwinter in vast roosts on the forest trees. Those from western populations, from the Rocky Mountains westward, move south to overwinter in southern California. The butterfly can avoid large bodies of water by flying low and accumulating along the windward shore, but over-water flight sometimes happens with a following wind. For example, monarchs from the eastern population are sometimes taken by north-west winds across the eastern coast of the USA, to reach Bermuda and even Europe (page 130). Others reach Florida or the Bahamas and then continue south-westward to Central America by overflying Cuba to the Yucatan peninsula (Urquhart & Urquhart 1979; see Fig. 51). During these long-distance movements, monarchs fly low, usually within 5 m of the ground at about 8 m/s and at temperatures greater than 10°C. They settle in vast numbers

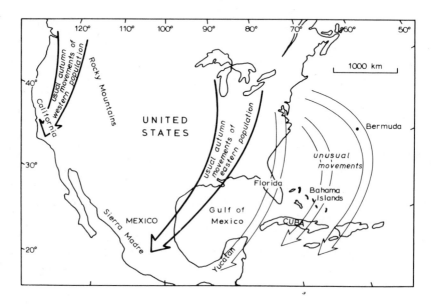

Fig. 51 – Generalised autumn movements of the monarch butterfly, *Danaus plexippus*. (After Urquhart & Urquhart 1977, 1978).

to avoid strong winds, and roost at night, particularly on cool nights following the southward passage of cold fronts. Flight is confined to sunny weather, and is strongly in straight lines. Observations on 477 individuals in the eastern USA showed that they flew singly, on slightly different tracks (Fig. 52), but strongly to the south-west (Schmidt–Koenig 1978), close to the ground in a head wind, but higher with a tail wind. Some field experiments with wild, autumn-migrating monarchs demonstrated the role of environmental cues in influencing direction of movement (Kanze 1977). Samples of 10–20 specimens were put in 80 cm-diameter circular, rotatable cages, each with a 20 cm-high periphery (either transparent or opaque) and with tops made of wire screening. The cages were placed in the middle of a field so that the butterflies could see the sky, but the (symmetrical) horizon only from the transparent cages. After 15 min, when most specimens were resting, the cages were rotated. It was found that the

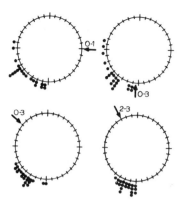

Fig. 52 – Four samples showing vanishing bearings of individual monarch butter-flies, *Danaus plexippus,* concentrated strongly towards the south-west despite differences in wind direction (speeds in m/s). Observations at Ithaca, New York. (After Schmidt–Koenig 1978).

butterflies sun-oriented, but only when it was shining, whether or not the horizon was visible. It seems that the monarch uses the sun to navigate during its autumn migration. Bearing in mind the thousands of kilometres covered during these migrations, together with the occurrence of spells of head winds, it seems that monarchs have insufficient fuel to complete the journey under their own power. Field observations show that the butterflies make extensive use of a variety of energy-saving flying techniques (Gibo & Pallett 1979), including thermal soaring in the presence of daytime convection (page 49) when there is a tail wind. Soaring is sometimes in circles, like birds, and individuals have been seen by glider pilots at heights above 1,000 m (Gibo 1981a). Another long-distance migrant that sometimes soars is the **mourning cloak butterfly,** or **Camberwell beauty,** *Nymphalis antiopa* (Gibo 1981b).

The **great southern white butterfly,** *Ascia monuste,* is another species whose boundary-layer migration has been well studied (Nielsen 1961). It is a tropical species, and its northern limit in Florida is set by the occurrence of damaging frosts. In the extreme south it breeds throughout the year in well-defined colonies on the coast, where it can find *Batis maritima,* the food plant of its caterpillar. Migration occurs between breeding sites by means of northward or southward straight flights in narrow streams, up to 10–15 m across, in both directions simultaneously. Flight is at heights of 1–3 m, with air speed varying with wind to keep a ground speed of 3–4 m/s. In light winds, flight is often along the tops of coastal dunes; in cross winds, flight is in the lee of vegetation. Migration stops in overcast weather, but continues when the sun is only tempor-arily covered by a cloud patch. This suggests that polarised light from the sky is being used for navigation, although visual orientation towards a goal seems to be involved, for when southward-flying butterflies left the southern tip of a very narrow north–south island, beyond which there was a stretch of about 500 m of open water to the next land to the south (Fig. 53a), they took a curved track, reaching the land from the west as though they were continually adjusting their heading towards the goal whilst being unable to correct for drift in the east wind over the open water.

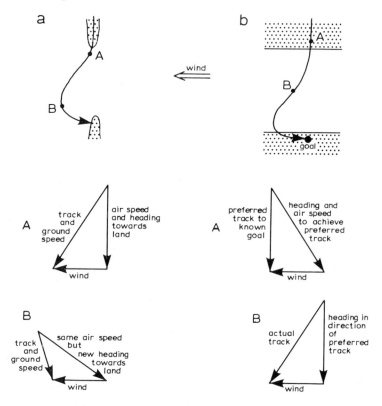

Fig. 53 – Insect flight over open water.
a   Great southern white butterfly, *Ascia monuste,* flying from one island to another, constantly changing its heading towards distant land – for example, southward at *A*, and south-eastward at *B*. (After Nielsen 1961).
b   Honeybee, *Apis mellifera,* flying across a river, showing rapid adjustment of heading from *A*, one that achieves the preferred track (using sky light patterns and allowing for wind drift over land) to *B*, one that substitutes preferred track for heading in the direction of the goal, and that therefore leads to downwind drift until the far bank is reached. (After von Frisch 1967).

The **Gulf fritillary,** *Agraulis vanillae,* is another butterfly whose autumn migration has been studied in Florida (Arbogast 1966). Numbered stakes were set out in a 15 m diameter circle at 20° intervals in a large field. Individuals crossing the circle were followed by noting the pairs of stakes nearest the places of entry and exit. Flight was at a height of about 1-2 m, and mostly towards 110-160°, with the wind being allowed for. Movement independent of changes in the synoptic-scale wind field has been further demonstrated by daily trapping of this and three other butterfly species (Walker & Riordan 1981). A similar compass orientation independent of wind direction was also found in experiments with tethered **large yellow underwing moths,** *Noctua pronuba,* flying at night in light winds and presumably within their boundary layer (Sotthibandu & Baker 1979). On moonlit nights, heading was related to the moon's bearing. If heading of any night-flying insect were dominantly towards the moon (Fig. 54), average track would be towards *west* of south (in the northern hemisphere) for a *waxing* moon (which is above the horizon longer before midnight than after), but towards *east* of south for a *waning* moon (which is above the horizon longer after midnight than before).

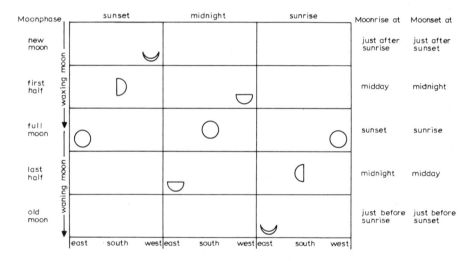

Fig. 54 – Variation of the moon's bearing with phase and time of night, at equinox in the northern hemisphere.

The **bogong moth,** *Agrotis infusa,* also makes seasonal migrations, presumably within its boundary layer. In spring, adults leave the winter pastures of eastern New South Wales as their food plant, *Medicago,* dies out, and they fly to the Australian Alps, where they gather as dense masses in crevices and small caves at or near summits higher than about 1300 m above sea level (Common 1954). After oversummering, they return northward, but clouds of moths sometimes appear on the east coast, presumably when west winds are strong. Another species, the **army cutworm moth,** *Chorizagrotis auxiliaris,* is widespread over the Great Plains of North America within 500 km east of the Rocky Mountains, from Canada to Mexico. In the spring, moths leave the plains and they cannot oversummer there. The newly-emerged moths fly westward to the high mountains, build up fat reserves, and then fly in autumn back to the plains, where they lay their eggs (Pruess 1967; Fig. 55). Any role played by the wind in this seasonal movement is not known. The same applies to the **larch bud moth,** *Zeiraphera diniana,* whose summer flights, apparently over hundreds of kilometres, lead to remarkably simultaneous outbreaks of caterpillars that cause defoliation in Alpine forests (Baltensweiler & Fischlin 1979).

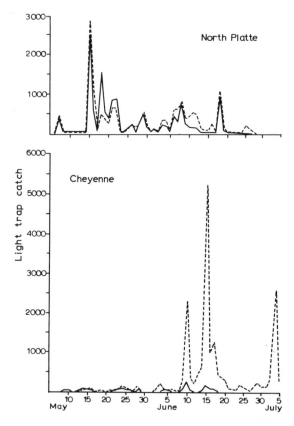

Fig. 55 – Daily light-trap catches of army cutworm moths, *Chorizagrotis auxiliaris,* in 1964, showing greater abundance and earlier peak of newly-emerged moths (continuous lines) at North Platte, Nebraska, compared with Cheyenne, Wyoming. where intermediate-aged moths were dominant (broken lines). Consistent with westward movement of the population. (After Pruess 1967).

Dragonflies are also strong boundary-layer fliers. For example, many *Hemianax ephippiger* and *Tramea basilaris* were seen flying northward close to the coast of north-west Africa, between Dakar in Senegal and Nouakchott in Mauritania, from 24 to 27 January 1976, against the usual north wind (Dumont 1976; see also page 142). Again, large numbers of *Pantala flavescens* appeared at Mugaga, near Nairobi, Kenya, on 12 October 1969, flying strongly from the south-west in east winds, and at heights mostly below 3 m (Odiyo 1973; see also page 129).

Honeybees, *Apis mellifera,* have an airspeed of about 8 m/s. They use ultra-violet and visible sunlight, and its pattern of polarisation from cloudless parts of the sky, to fly from hive to food source; and they change heading to compensate for drift in the presence of wind, but they cannot do that over open water (von Frisch 1967, see Fig. 53b). Queen **bumble bees** are often seen on warm days in spring along the south-west coast of Finland, up to 900 an hour, flying south-eastward against the wind and towards Russia (Mikkola 1978). Flight is mostly below 5 m, but below 1 m in windy weather.

Free flight at night can be watched with radar (page 163). On 3 November 1973 at Kara, Mali, insects were seen by radar to fly against a wind of 2 m/s and at a ground speed of 3 m/s (that is, with an air speed of 5 m/s) (Riley 1975). They were possibly various species of **grasshoppers**, flying with a common heading at heights of several hundred metres above the ground and about 50 m apart.

4.2 UPWIND FLIGHT

In contrast to this concerted and persistent flight, boundary layer flight can be much more variable and intermittent, as individuals seek particular goals such as food, shelter, mates or egg-laying sites. In many cases, flight is *upwind,* probably within an odour plume. (For a broader discussion of the effects of odours on insect behaviour see, for example, Shorey & McKelvey 1977.) **Tsetse flies,** *Glossina morsitans,* and *G. pallidipes,* were caught coming to an ox in woodland at Rekomitjie, Zimbabwe. They were mostly unfed females, approaching upwind from distances out to 90 m (Vale 1977; see Fig. 56). When not flying upwind in the odour plume from the ox, the flies were found to be ranging (Vale 1980). Similar flights are made by **mosquitoes** seeking a blood meal. Stereophotography of the **yellow fever mosquito,** *Aedes aegypti,* in a wind tunnel has shown that its flight is in short, straight segments, the frequency distribution of durations being logarithmic, implying an absence of external triggers and a tendency to turn at random. In a plume of air blowing across a human hand, however, the turning pattern changed on leaving the plume (but not on entering), with much sharper turns that often led to re-entering the plume (Daykin *et al.* 1965). It was also shown that this species needs visual contact with the ground to make an upwind flight, for there is no concerted heading in darkness. A similar conclusion was reached with flight studies of the **vinegar fly,** *Drosophila melanogaster* (Kellogg *et al.* 1962), and so did the moth *Hadena bicruris,* flying in the scent from flowers (Brantjes 1981). In similar experiments with the **stable fly,** *Stomoxys*

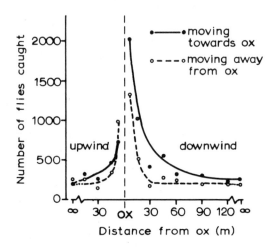

Fig. 56 — Numbers of the unfed tsetse flies. *Glossina morsitans* and *G. palli-dipes* caught in nets at various distances upwind and downwind of a tethered ox in woodland at Rokomitjie, Zimbabwe, showing approach from the downwind side. (After Vale 1977).

calcitrans, a widely distributed blood-feeding fly that is a pest of livestock, reducing milk yield and possibly carrying disease organisms, Gatehouse & Lewis (1973) found that a combination of carbon dioxide and air that had crossed a human hand led to upwind flight, as it did with *A. aegypti* (Mayer & James 1968), as well as with *Anopheles arabiensis* and *Culex pipiens fatigans* (Omer 1979). Field experiments in the Gambia (Gillies & Wilkes 1969, 1970, 1972) have shown that several species of culicine **mosquitoes** were attracted by carbon dioxide from distances up to 20–40 m (Fig. 57). Some anopheline species, by contrast, including *Anopheles melas,* a **malaria** carrier, were attracted from some-what greater ranges by odour from calf or human bait. For two *Mansonia* species there was evidence that newly-emerged females were less responsive than older ones, as would be expected if initial flights were ranging, or even above the boundary layer, rather than upwind towards a food source. Catch was also found to decrease to zero as wind speed approached the air speed of host-seeking females (Gillies & Wilkes 1981). Similarly, the **hawkmoth** *Deilephila elpenor* has been shown in wind tunnel experiments to be unable to reach flowers by means of their scent plumes when the wind speed exceeded the moth's air speed of about 5 m/s (Brantjes 1981).

A possible mechanism by which a night-flying insect can judge the wind direction without the aid of visual cues (including apparent movement of the ground) has been suggested by Gillett (1979). In an airstream with strong vertical wind shear (page 38), such as is common at night near the ground, a sudden up or down flight will lead to a change in ground speed, but momentarily

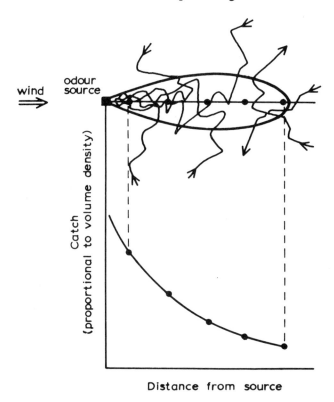

Fig. 57 – Preferential flight within an odour plume leads to greater catches towards the source. Plan of *time-averaged* boundary of that part of the plume where newly-emtered insects have a given chance of reaching the source.

there is a lag during which the forward *air* speed will be different from what it was just before the vertical movement. For an upward move, the air speed will be momentarily *less* for an insect heading *down*wind, but *greater* for one heading *up*wind (Fig. 58). The reverse will be true for a downward move. Thus, by detecting momentary changes in the air speed as it flies suddenly up or down, an insect can tell if it is flying upwind or downwind. It is not known how widely such a mechanism is used, if at all.

Field studies of **ambrosia beetles,** *Trypodendron lineatum,* in British Columbia showed that they flew upwind to cut logs. By hanging a white cloth parallel to the wind it was found that the flux of beetles passing the cloth decreased sharply either with very light winds or with winds stronger than about 1.5 m/s (Fig. 59) – presumably because in the lightest winds the odour plume is ill-defined, whereas on the strongest winds the beetles could make no headway because of their small air speed (Chapman 1962). The **mountain pine beetle,** *Dendroctonus ponderosae,* similarly flies upwind in the presence of an attractant

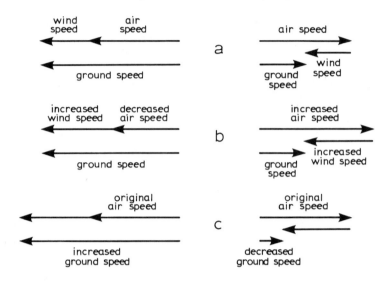

Fig. 58 – A mechanism by which a night-flying insect might judge wind direction in the presence of positive vertical wind shear. When it climbs suddenly, *ground* speed tends not to change immediately, due to conservation of momentum; hence *air* speed changes as wind speed increases (b). Continual flapping, however, soon restores the air speed (c), and hence changes the ground speed. The insect temporarily experiences a *reduced* air speed if it is heading (at least partly) *down-wind* (b of the left side) but an *increased* air speed if it is heading (at least partly) *upwind* (b of the right side). (In the same way, an aircraft can stall if its air speed decreases too much as it descends into a stronger tail wind. This has been a cause of crash landings in outward-spreading squalls from rainstorms – page 50). (After Gillett 1979).

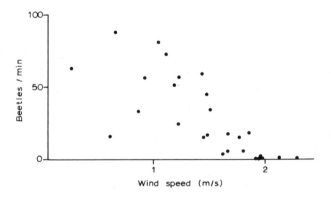

Fig. 59 – Flow rate of ambrosia beetles, *Trypodendron lineatum,* passing a given point, in relation to wind speed. (After Chapman 1962).

(Gray *et al.* 1972), and it is likely that newly emerged *Scolytus* beetles, the carrier of spores of the fungus causing **Dutch elm disease,** fly to feeding places in twig crotches of healthy host trees by means of odour plumes, for they are not attracted to similar twigs of other tree species (Keyserlingk 1980). Field studies of the **green lacewing,** *Chrysoperla carnea,* in California showed that after marking, release, flight and landing, they took off again upwind to an artificial food source, and seldom at heights above 1 m (Duelli 1980). Wind tunnel experiments with the Hawaiian **fruit fly,** *Drosophila mimica* showed that it moved upwind when wind speeds were less than about 1 m/s, but downwind in stronger winds (Richardson & Johnston 1975). These results agreed with movements in the field by flies marked with tracer chemicals.

Upwind flight by moths to food plants of their caterpillars has been demonstrated with the **navel orangeworm moth,** *Amyelois transitella.* About 1600 moths from laboratory caterpillars fed with a red dye, so that they laid pink eggs, were released in a 60 ha almond orchard in California (Andrews *et al.* 1980). From the pattern of pink eggs in a network of egg-laying traps out to 300 m it was found that females flew upwind (Fig. 60). By following individual egg-laying **European pine shoot moths,** *Rhyacionia buoliana,* as they flew from tree to tree in a heavily infested plantation, Green & Pointing (1962) found

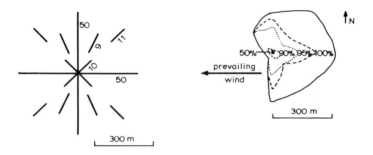

Fig. 60 – Distribution (right) of marked eggs of the navel orangeworm moth, *Amyelois transitella,* laid in a network of traps arranged in linear groups as shown (left), within a grove at McFarland, California, and showing upwind flight of gravid moths seeking laying sites. (After Andrews *et al.* 1980).

that when winds were less than the moth's air speed (about 5.5 km/h) flight was upwind. The **rice bug,** *Leptocorisa chinensis* also flies upwind towards the source of rice plant odour (Kainoh *et al.* 1980), and so does the **onion fly,** *Hylemya antiqua.* In field observations with sliced onion, *Allium,* barnyard grass, *Echinochloa,* and bare soil, more flies were seen to land within 0.5 m of onion than with other treatments (Dindonis & Miller 1981). Both males and females flew upwind in short steps with landings. When landing downwind of onion they turned into wind and then flew straight to the odour source. Similar flight was found in a wind tunnel with gravid female **cabbage root fly**, *Delia brassicae* (Hawkes *et al.* 1978), and this kind of flight may be a common way by which weak fliers can reach an odour source. With this species, only mated, gravid females flew towards host plants (Hawkes & Coaker 1976). Caged individuals tended to gather on the upwind side where cages were placed downwind of host plants (Hawkes 1974), the greatest response distance presumably depending on source strength, wind speed and gustiness. By marking with ^{32}P, upwind flight was demonstrated in the field (Fig. 61). Experiments in Sweden with **brassica pod midges**, *Dasyneura brassicae*, also marked with ^{32}P and released on a cereal crop showed that on some days females moved *upwind within* the crop towards host plants; whereas *above* the crop, movement was *downwind*, presumably reflecting the differences in flight within and above the insect's boundary layer (Sylvén 1970). Similar flights are made by female **wheat blossom midges**, *Contarinia tritici.* coming from their emergence sites (Basedow 1977).

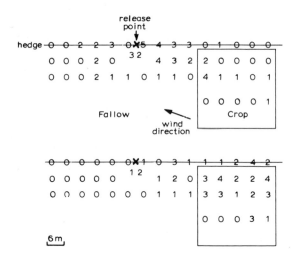

Fig. 61 – Distribution of female cabbage root flies, *Delia brassicae*, marked with ^{32}P, and caught in a network of traps, showing upwind flight into a crop. 1,930 flies released at 2300 LT at the point shown on the edge of the field. Catches on the following day: 0430–1130 (above) and 1130–1830 (below). (After Hawkes 1974).

An interesting effect of boundary-layer flight controlling the effectiveness of a parasite comes from Hawaii, where the ichneumonid, *Trathala flavoorbitalis* is the main larval parasite of the **coconut leaf roller**, *Hedylepta blackburni*, a moth pest of coconut palm, *Cocos nucifera*, on the windward side of the islands. Because of the wind strength, this parasite is unable to reach the most windward trees in numbers enough to act as a natural control, which it is in the more sheltered trees (Bess 1974).

It has long been thought that insects used scents for mate finding. The fluttering of males around a caged female *Biston betularia* was recorded 300 years ago, and collectors have used caged females to lure males of rare species (Inscoe 1977). By 1920 it had been found that extracts of the abdominal tips of females could be used in place of live insects, but it was not until 1959 that the first clear identification of the scent had been made — that of the **silkworm moth,** *Bombyx mori.* Substances such as this, emitted by one individual to elicit a specific response in another of the same species, are known as **pheromones.**

How an insect finds a pheromone source is still not well understood. Flight may be thought of as being at first downwind or crosswind until a sought-after plume is found and the odour strength is greater than a threshold that turns flight upwind. Because sex pheromones are often mixtures of closely similar chemicals, the threshold concentration may well vary with the proportions present (Roelofs 1978; see Fig. 62). The stronger the source, the greater is the range likely to be over which upwind flight is triggered. Thus, by using caged male **oriental fruit moths,** *Grapholitha molesta,* Baker & Roelofs (1981) showed that the distances downwind at which they started upwind flight about doubles for a tenfold increase in emission rate (as would be expected if plume concentration is inversely proportional to the cube of the distance). They also found that flight ended further from the source with greater emission rate, with a three orders of magnitude difference in calculated pheromone concentration between the apparent upper and lower thresholds. By using a night viewing device, Lingren *et al.* (1978) found that male **tobacco budworm moths,** *Heliothis virescens,* and **cabbage looper moths,** *Trichoplusia ni,* flew rapidly crosswind for about an hour before mating was seen or males were caught in pheromone traps.

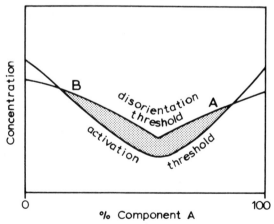

Fig. 62 — Suggested schematic variation of threshold concentrations with composition of pheromone having two components, *A* and *B*. The intersecting curves define the shaded space within which pheromone will induce upwind flight. Curve shapes probably vary with the chemical nature of the pheromone. (After Roelofs 1978).

Once flight turns upwind along the plume, observations from several species suggest that it takes place in three stages (Fig. 63; see also Kennedy (1978) and Mankin *et al.* (1980)):

(a) furthest from the source, where the plume is wide and weak, flight is fairly fast and more or less straight upwind;
(b) nearer the source, where the plume is stronger but narrower and more easily lost, flight is slower and in a zig-zag path, still overall upwind but with turning at the edges of the plume (sensed maybe by changing smell strength);
(c) close to the source, where the plume is very narrow and strong, flight is slower still, the path is zig-zag, and changes in flight become triggered by what the insect sees or hears, rather than smells, perhaps because another, but higher, threshold is exceeded.

Fig. 63 – Schematic flight track (thick line) within a pheromone plume that is formed of innumerable turbulent contortions of the filament initially drifting from the source (thin line).

This kind of flight pattern may be more typical of strong fliers, contrasting with the intermittent flight of some weaker fliers (see page 105).

If the plume is lost, flight is turned crosswind, but if the plume is not regained, flight is turned downwind (Traynier 1968). *Drosophila melanogaster,* flying in an odour plume from fermenting bananas, behaved very similarly (Kellogg *et al.* 1962). Using infra-red photography in Crete, Murlis & Bettany (1977) found that flight of *Spodoptera littoralis* males towards a synthetic pheromone source was very consistent (Fig. 64). Beyond about 2–4 m, flight was straight and level at a height of 0.2–2 m and at a ground speed of 3 m/s. Nearer the source, flight zig-zagged in the vertical plane, and within about the last 50 cm ground speed fell to zero, together with vertical zig-zags whilst still heading upwind. It seems that final location of a calling female depends on sight, or sense other than smell.

Kennedy *et al.* (1980, 1981) have suggested an alternative mechanism for reaching a scent source: upwind flight is triggered by detection not just of the mere presence of the pheromone but of increases and decreases in concentration. And further, they suggest that crosswind tacking is a response not to loss of or decreased concentration of pheromone but to presence of a uniform concentration of pheromone. Wind tunnel experiments with male **summerfruit tortrix**

Fig. 64 – Track of a male cotton leafworm moth. *Spodoptera littoralis,* flying freely at night towards a pheromone source. Photograph by infra-red light and multiple exposure at 1/30 second intervals. Strong reflection from the source's mounting causes the highly distorted image on the right. Distant village lights appear in the background. Drapanias, Crete, 1977. (After Murlis & Bettany 1977).

moths, *Adoxophyes orana,* show this. One side of the tunnel was given a uniform pheromone cloud whereas the other was kept clean except for a plume that could be removed at will. The cloud had a sharp boundary, as shown by smoke simulation. Males advancing in zig-zag flight up the plume began casting from side to side when the plume was suddenly removed. When this casting led to entry into the uniform cloud there was a short-lived resumption of upwind flight (so long as the moths had spent at least 2 s in clean air), followed by further crosswind casting. On finding a plume embedded within the cloud, upwind zig-zag flight was resumed, suggesting that although a uniform cloud soon lost its ability to trigger upwind flight, sudden increases restored it. Entry into cloud from clean air led to smaller crosswind movements that persisted for at least 2 s after re-entering clean air. As a result, casting was concentrated along the cloud boundary, and track reversal often occurred within the cloud and therefore could not have been triggered by loss of scent.

Different mechanisms may well be used by different species, for wind tunnel experiments with the **Mediterranean fruit fly**, *Ceratitis capitata,* have shown that it flies straight upwind in a uniform odour, but in zig-zags in a plume (Jones *et al.* 1981).

Caged male **cabbage looper moths,** *Trichoplusia ni,* increase their wing fanning and upwind movements towards the scent source when as much as 50 m downwind of a synthetic sex pheromone source (Kishaba *et al.* 1970). Some marked and released males were recovered at the source within 15 min, the proportion decreasing with increasing distance from the release point. In a 10 ha field of 1.3 m high cotton in California, Kaae & Shorey (1973) set up phero-mone traps at 0.3 m and at crop top to catch **pink bollworm moths,** Pectino-phora gossypiella. The ratio of catch sizes, bottom/top, increased with wind speed (Fig. 65). In light winds, the plume is presumably most effective in the upper canopy, for there would be little wind near the ground. By contrast, a strong wind would not allow upwind movement, except in the lower canopy.

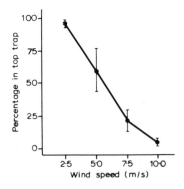

Fig. 65 – Influence of wind speed on catches of male pink bollworm moths, *Pectinophora gossypiella,* expressed as ratios of catches in traps at cotton top and at 0.3 m above the ground. (After Kaae & Shorey 1973).

Two ways have been put forward whereby it is thought that a flying insect uses a pheromone plume to find a mate (Farkas & Shorey 1974): **chemotaxis,** using plume structure, and **anemotaxis,** using the wind.

Chemotaxis assumes that the contorted thread-like structure of the plume provides information about the direction and distance of the source by means of smell pulses with intensities and spacings that may vary with distance from the source. There is no clear evidence, however, that insects use this method (Marsh *et al.* 1978). Anemotaxis assumes that the insect turns to fly a track against the wind, or at some known angle to the wind (**anemomenotaxis**). The insect needs to know its track in relation to wind direction, and it needs to be able to maintain it. Evidence from both field and laboratory suggests that anemotaxis is used by at least several kinds of insects (mosquitoes, honeybees, fruit flies, aphids, moths) to reach pheromone sources (Kennedy 1977). Wind tunnel experiments with the moth *Plodia interpunctella* showed that in its zig-zag flight, both track angles and track reversal frequency stayed constant, even though distances between reversals decreased (that is, there was a narrowing of the zig-zags) as the moth homed in on the source (Marsh *et al.* 1978, 1981). This was achieved by flying slower and turning more into the wind. Sudden removal of pheromone plume caused little change in pattern of zig-zag flight for a short while (less than a second), as might happen when over-shooting a source, or when flying in a patchy plume, but it was followed by wider cross-wind casting flights (that is, track angles and reversal frequency both increased due to a turning away from the wind). Nor did changing the wind speed alter the track angle and reversal frequency, because the moths flew faster and turned more into wind. It is supposed that these manoeuvres are cued by movement of ground images across the eyes, modulated by measurement of air speed. Kennedy (1939) showed this with wind tunnel flights of the **yellow fever mosquito,** *Aedes aegypti,* and no other mechanism seems able to explain upwind orientation. How the central nervous system integrates the available information is not yet known. Using a wind tunnel with a patterned, moveable floor, and the moth *Plodia interpunctella,* Kennedy & Marsh (1974) found that flight speed diminished in accordance with increasing floor speed, from which it can be inferred that the moth responds to the ground pattern. Moths with only one antenna flew in a way indistinguishable from normal moths, hence it does not seem that moths use lateral odour gradients such as might be detected by paired antennae. Cardé & Hagaman (1979) found that as the strength of the source of an artificial pheromone, disparlure, was increased so both the upwind ground speed and the extent of cross-wind zig-zagging decreased. This suggests that moths use not only the *presence* of the plume to indicate *direction* of the source, but also the *intensity* of odour to indicate *distance* of the source. It is not only moths that seem to use positive optometer anemotaxis, for males of the sciarid **fly** *Bradysia impatiens* show similar zig-zag flight to a source of female scent (Alberts *et al.* 1981).

It is not yet possible to measure the instantaneous structure of a plume, with its minute quantities of pheromone. Close to the source it is likely to resemble a thread with a width less than the size of the smallest atmospheric eddies. Downwind, the thread becomes much contorted by eddy mixing, which eventually leads to a broader but still patchy plume, rather like the smoke rising from a cigarette burning in still air (Fig. 63; and see page 184 for further discussion of plume dispersion). Objects in the way of the wind will further alter plume shape. Murlis & Jones (1981) have simulated plumes by using ionised air. Field observations up to 15 m downwind of an ion generator suggest that odour arrives in discrete bursts (Fig. 66), but neither strength nor return period seem to vary with distance from source, and peak concentration is not a reliable guide to distance from source. Distance might be estimated by an insect, however, if it could average concentration over many seconds. Effective plume length will vary with source strength and time, with wind speed and gustiness, and with the insects' air speed and ability to smell. It is usually some tens or hundreds of

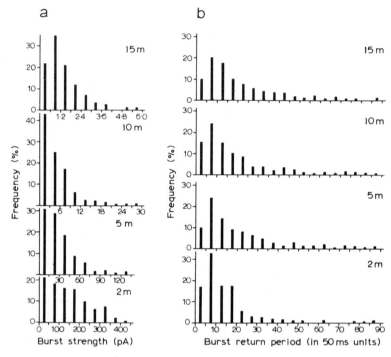

Fig. 66 — Frequency distributions of burst lengths (a) and burst return periods (b) in a stream of ionised air 15 m downwind of a source, and stimulating the pulses to be expected in a drifting pheromone plume with a structure like that of Fig. 63. (After Murlis & Jones 1981).

metres, as already noted, but in some species may be a few kilometres. Early calculations of plume length (for example, Sower *et al.* 1973), based on diffusion theory (Bossert & Wilson 1963), are misleading in that they assume *time-averaged* plumes rather than highly filamentous, instantaneous ones. Moreover, plume structure is likely to vary with the degree of atmospheric gustiness and mixing. Early and late in the day, and at night, plumes are likely to be narrower than during the middle of the day. Such differences may account, in part, for differences in flight from one day to another, and between one part of a day and another (Chapman 1967). Field observations with the **tobacco cutworm moth,** *Spodoptera litura,* gave about 60 m as the greatest downwind effectiveness of its pheromone plume (Nakamura & Kawasaki 1977). Other observations in Cyprus with a related species, the **cotton leafworm moth,** *Spodoptera littoralis,* using virgin female traps that recorded the exact arrival time of males, showed that catch increased with wind speed, perhaps because females released pheromone for longer periods in stronger winds (Campion *et al.* 1974). In Israel, using the same species marked by means of a red dye in their food, Kehat *et al.* (1976) found that the effective range was about 100 m.

From experiments with networks of traps baited with sex pheromone of the **pea moth,** *Cydia nigricana,* it is clear that wind speed and direction affect catch, as well as trap spacing (Wall & Perry 1980). Trap catch decreases to zero as wind speed approaches the insect's air speed. This has been found, for example, with **red bollworm moths,** *Diparopsis castanea,* in Malawi (Marks 1977), and with **tobacco budworm moths,** *Heliothis virescens* and **cabbage looper moths,** *Trichoplusia ni* in the USA (Hendricks *et al.* 1980).

Sex pheromones have been used in population management in three ways — monitoring, suppression, and disruption of communication (Cardé 1976, Inscoe 1977). As early as 1914, females of the **gypsy moth,** *Lymantria dispar,* were being used in forests to find new infestations. Monitoring is now being used for several pest species to find the most effective time to apply insecticide. Suppression by using traps to compete with females has so far been successful with only a few species, but even with those the method is economically impractical. Disruption of mating was first tried in 1967, and is proving to be successful with a growing number of species. Males may be confused by an atmosphere permeated with synthetic pheromone, so that they cannot find wild females, or the threshold concentration that triggers upwind flight may be raised by the persistent presence of the synthetic, so that the calling range of wild females is reduced. Disruption can be by many point sources or by microencapsulation, whereby the whole area treated becomes effectively one large source.

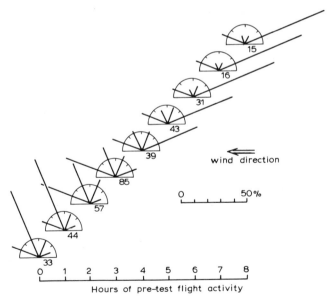

Fig. 67 — Percentage distribution, among four sectors in a vertical plane, of flight tracks of freshly-emerged, unfed male European elm bark beetles, *Scolytus multistriatus,* taking off in pheromone-permeated air in a wind tunnel, in relation to time spent flying before testing. (After Choudhury & Kennedy 1980).

Effective monitoring or disruption by point sources depends upon good trap design and positioning. Lewis & Macaulay (1976) have shown that traps giving elongated plumes (as simulated by smoke) and placed at heights where there is still some wind but less than the air speed of the insect, had the greatest catches of the **pea moth**, *Cydia nigricana*.

Pheromones for aggregation, not mating, can also induce positive anemotaxis. Newly-emerged bark beetles (Scolytidae) fly briefly to host trees and release aggregation pheromones which, together with host volatiles, attract other beetles and lead to mass attack. Great economic damage is done in forests as a result, especially in coniferous forests of the northern hemisphere. **European elm bark beetles**, *Scolytus multistriatus,* that had flown for more than 6 h in a wind tunnel in the presence of diffused pheromone subsequently flew mostly upwind (Fig. 67), presumably in an attempt to reach the apparent food source, as if earlier arrivals had been releasing the pheromone (Choudhury & Kennedy 1980).

CHAPTER 5

Insect flight above the boundary layer

Because a *swarm* of insects is conspicuous, it can be tracked for tens or hundreds of kilometres across country (page 169). By contrast, it is not yet possible to follow an *individual insect* for more than a few hundred metres. The most direct evidence for windborne migration of an individual insect over distances of tens or hundreds of kilometres, or more, comes from the possibility of back-tracking between its place of sighting and its place of origin, using trajectories based on weather maps. The insect needs to be marked so that it can be distinguished from others of its species. Markers can be artificial (such as dye or radioactivity — and the source can then be known precisely) or natural (such as age, body shape, egg development, parasite load, gut content or chemical composition — and the source is then less definite). To calculate a back-track, the time and place of sighting need to be known as precisely as possible because the windfield varies with time. Until now, there have been few studies in which individual insects have been *both* marked *and* back-tracked using maps of windfields. On the other hand, there have been many studies *either* of unmarked but back-tracked insects, *or* of marked but untracked insects. These two kinds of evidence for windborne migration are both circumstantial; that from the back-tracking of unmarked individuals is more convincing than that from mark-and-recapture field experiments that have taken little account of possible effects of wind on movement.

There are other kinds of evidence suggesting that a given insect movement might be windborne. For example:

- sudden outbreaks of larvae or of insect-borne disease (of plants, animals or man) consistent with back-tracks of parents or vectors from known or likely sources;
- sudden mass arrivals or departures of insects in association with particular kinds of weather systems, even though no back-tracks have been calculated.

There have also been many studies of sudden arrivals of flying adults in

traps or in crops, especially first arrivals in areas that have been cleared artificially or that are unable to produce flying populations, but where no relationship to the wind has been looked for. *Simultaneous* arrivals at places far apart,
or arrivals far from known sources (for example, at sea or in deserts) are particularly convincing. Another kind of powerful circumstantial evidence in favour
of windborne migration comes from the biogeographical analysis of the seasonal
redistribution of the species in relation to seasonal wind changes. It requires
extensive field surveys of both insects and winds, preferably over many years.
Yet other kinds of circumstantial evidence are:

- flying behaviour of individuals in relation to the wind at some given
  time (by eye or by radar);
- flight characteristics of an individual (duration, height and speed) in
  relation to wind and distance moved.

It is clear that a great variety of evidence can be sought to suggest whether
or not a particular species might be windborne. The many examples that follow
have been grouped to illustrate this variety. Some species are mentioned under
several headings because they are pests of great economic importance. For other
species, mentioned perhaps only under one or two headings, further evidence
might be available already, or might be deduced from existing but unpublished
records. Evidence is mostly circumstantial. Each piece may not be very convincing on its own, but taken with other pieces there is little doubt that many
species can, and often do, move downwind, sometimes over great distances.

5.1 MARKING AND BACK-TRACKING

The results of field experiments with five species can be quoted to illustrate
the limited amount of work done so far on the movement of marked individuals
in relation to the wind. First, we consider the **cabbage root fly**, *Delia brassicae*.
When 1400 one-day-old, dye-marked males were released in the middle of a
13 ha field of flowering mowing grass, *Lolium perenne,* surrounded by a rec-
tangular grid of 145 yellow water traps at 14 m intervals, they were recaptured
downwind on each of three days following release in persistent north-east winds.
Their average displacement after three days was 155 m (Finch & Skinner 1975).
Such short distances imply flight above the boundary layer totalling no more
than a few minutes. In contrast, the distribution of the few recaught females, of
1100 released at the same time, was not related to the wind. This and other
evidence suggests that females generally left the trapped area and therefore
avoided recapture, whereas the few that remained near the point of release
presumably flew within their boundary layer and were therefore able to reach
traps in all directions.

About 15,000 pods of oil turnip that had been treated with ^{32}P were placed
at the centre of a 50 m X 50 m square of fallow, and the **brassica pod midges,**

Dasyneura brassicae, that emerged were sampled at 12 points along a 5 m wide strip of oil turnip crop surrounding the square. Of the 2241 radioactive midges caught during 6 days, 83% were female, and the movement was largely down-wind (Sylvén 1970; see Fig. 68). Such small insects, however, are weak fliers, and their boundary layer will be very shallow on many days.

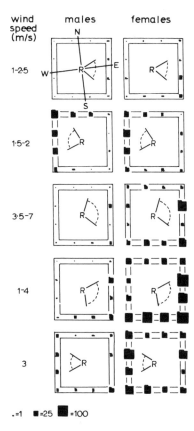

Fig. 68 – Examples of distributions of adult brassica pod midges, *Dasyneura brassicae,* marked with ^{32}P and sampled by sweep nets in a 5 m wide strip of summer oil turnip around a 50 × 50 m square of fallow at Bergshamra, Sweden. Winds measured at a height of about 1.5 m above the ground – speeds shown at far left, and approximate range of directions shown at centre of square, where the midges were released. Males left, females right. (After Sylvén 1970).

Moths of the **cotton leafworm,** *Spodoptera littoralis,* a pest of many crops in Africa and around the Mediterranean, were marked with red fluorescent dye and released at sunset from Abou Rawash, about 30 km south-west of Cairo, and recaptured within four hours at white screens held behind kerosene lamps placed at distances up to 1500 m in various directions from the release point

(Salama & Shoukry 1972). Most moths were recaught to the south and west, consistent with the dominance of north and north-east winds at the time. Two were caught at 1500 m. For experiments with males of this species flying in their boundary layer, see page 108. **Hop aphids,** *Phorodon humuli,* marked with 32P, were put into plum trees at Steknik, near Zatec in Czechoslovakia, six days before a mass flight (Taimr & Kriz 1978). Using 102 sticky yellow traps among 80 ha of hop gardens, and 69 places for collection on leaves, as well as direct radiometric measurement of the upper leaves of 15,000 hop plants up to 2 km from the source, marked aphids were found to the north-east, very likely due to a downwind movement on warm, light south-west winds on one particular day, contrasting with weather too cool for flight on other days with light winds. (See pages 157 and 166 for further evidence that this species is a windborne migrant.) In similar experiments in Canada with two species of **mosquitoes,** *Aedes sticticus* and *Aedes vexans* (whose larvae were collected, put in rearing boxes, and the adults separated and dusted with fluorescent dyes before release), individuals were recaught in light traps up to 11 km from the release point, and mostly downwind (Brust 1980; see Fig. 69).

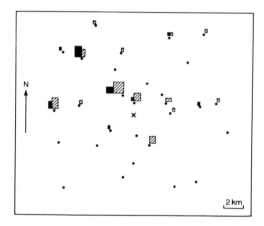

Fig. 69 – Distribution of 66 mosquitoes, *Aedes sticticus* and *A. vexans,* marked with fluorescent dyes, released at point x, and recaught in a network of 29 light traps at Winnipeg, Manitoba over a 19-night period with mean winds from south and south-east, and showing movements mainly downwind. Males (black squares) and females (hatched squares). The largest square represents 20 mosquitoes. (After Brust 1980).

These experiments are less convincing than they might have been if many more individuals had been marked, and if winds had been measured at the precise time of flight.

5.2 MARKING, BUT NOT BACK-TRACKING

Vast numbers of insects need to be marked and released if movements over distances of tens or hundreds of kilometres are to be demonstrated. This has been done in China with the moth of the **oriental armyworm,** *Mythimna separata,* a pest of cereals and a wide range of other plants over southern and eastern Asia. About two million moths were attracted to baited straw bundles during 1961-63, marked by spraying with an alcohol solution of dyestuff, released and later recaught with baited straw. Twelve were recovered after straight-line movements of 600-1400 km (Fig. 70), taking 1-3 weeks (Li *et al.* 1964). Other evidence (pages 137 and 144) strongly suggests that this species migrates on the wind. Similar experiments, marking millions of **brown planthoppers,** *Nilaparvata lugens,* have demonstrated movements over hundreds of kilometres in southeastern China (anon 1981). There is evidence (pages 134 and 144) that this species is a windborne migrant. There have been other experiments with moths in Texas, but on a smaller scale. About one million pupae of the **tobacco budworm moth,** *Heliothis virescens,* were released from a site about 30 km north-north-west of Brownsville over about four months in 1971 (Hendricks *et al.* 1973). The caterpillars had been fed in the laboratory on a diet with a red dye that was passed on to the new adults, and the pupae had been sterilised by radio-

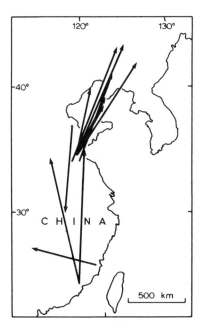

Fig. 70 – Arrows joining release points and recapture points of dyed oriental armyworm moths, *Mythimna separata,* in China, 1961-3, showing movements over many hundreds of kilometres, and taking 1-3 weeks. (After Li *et al.* 1964).

active ^{60}Co before release. Female-baited traps were set up on eight compass directions out to 100 km and changed about every six days. Eight males were caught at 100 km, as well as others nearer the release point, and there was a general trend for most moths to be caught downwind from the release point. Further experiments with this same species in the West Indies have demonstrated inter-island movements of at least 80 km (Haile *et al.* 1975; see Fig. 71). Sixty thousand dyed and sterilised, laboratory-bred pupae were released daily for 32 days on the island of St Croix (18°N 65°W). Eighteen emerged male moths were caught in female-baited traps on the neighbouring island of Vieques, 80 km to the north-west, and five in sticky traps on St Thomas, 60 km to the north. In similar experiments with the **corn earworm moth,** *Heliothis zea,* using 6,000 pupae a day, two moths were caught on Vieques. There is evidence that both these species are windborne migrants — see pages 154 and 155 for *H. virescens,* and pages 153 and 155 for *H. zea.*

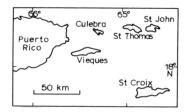

Fig. 71 — Location map showing places where dyed tobacco budworm moths, *Heliothis virescens,* and corn earworm moths, *H. zea,* were recaught after release on St Croix. See text.

The **screw-worm** is the maggot of a fly, *Cochliomyia hominivorax,* that lays its eggs in the wounds of live, warm-blooded animals, and the maggots feed on animal tissue. It is a western hemisphere species of tropical and subtropical latitudes, being killed by frost. In 1959 it was eradicated from its overwintering sites in southern Florida (Knipling 1960) by releasing millions of sterile males. Nevertheless, in the following two years a few flies were caught in northern Florida and many more further west, implying invasion from south-western USA (Baumhover 1966), but an eradication programme there became progressively more effective until the USA was free and sterile male releases could be confined to a strip along the border with Mexico. This species is clearly very mobile, and dyed, sterilised flies have been released in Texas to measure flight range. Each week for 8 months in 1963, 200,000 laboratory flies were released from the same site. Flies were caught again out to nearly 300 km (Hightower *et al.* 1965). It seems likely, but it has not been demonstrated, that flies are taken downwind. Nellis (1977) has proposed that some Caribbean islands were invaded when spells of west winds temporarily interupted the dominant easterly trade winds, and that flies were taken from Puerto Rico, before the successful eradication programme there in 1971–4.

A chance marking of an individual moth has also demonstrated movement from North Africa to England (Kettlewell & Heard 1961). A specimen of *Nomophila noctuella* collected on 10 March 1960 was found to be carrying a 9 μm diameter radioactive particle consistent in activity and appearance with such particles derived from the French atomic weapons test on 13 February in southern Algeria. The radioactive cloud from the test moved east (judged from wind-based trajectories) and did not reach western Europe until 26–29 February, after passing around the world. Because it is unlikely that a 9 μm diameter particle would have stayed airborne until then, it seems to have been picked up by the moth in Africa, therefore giving evidence for a movement of about 2,000 km.

Three million dye-marked **melon flies**, *Dacus cucurbitae*, were released as sterilised pupae each week for two years on Kume Island in the Okinawa group as part of an eradication scheme (Kawai 1978; see Fig. 72). During May and June, 15 marked flies were caught (out of a total of 385) on downwind islets up to 50 km away, but none was marked among the 3073 caught in Okinawa, 100 km away. It seems that flies were carried on the frequent west winds, for they would have been unable to fly 50 km unaided. On another occasion, one fly released on Kume was recaptured about 200 km to the north-east, on Okinoerabu Island, but because the trap there was emptied only every two weeks the date of movement, and hence its association with winds at the time, cannot be determined (Miyahara & Kawai 1979). Some hundreds of thousands of paint-spotted **boll weevils**, *Anthonomus grandis,* were released in the USA within large networks of pheromone traps. About 50 were recaught for every 100,000 released, the greatest distances moved being about 50 and 70 km in two sets of experiments (Johnson *et al.* 1975, 1976). There is other evidence that this weevil moves long distances on the wind (page 153).

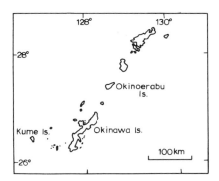

Fig. 72 – Location map showing places where dyed melon flies, *Dacus cucurbitae,* were recaught after release on Kume Island. See text.

Mark-and-recapture has been used to demonstrate migration in West Africa by scattered **African migratory locusts,** *Locusta migratoria, migratorioides,* moving as individuals and not as swarms (page 178). Towards the end of 1955, after the rainy season, and again in 1956, Davey (1959) marked about 200,000 locusts in the semi-arid country around the flood plains of the River Niger. Marked locusts were first caught in the flood plains at the southern end, where floods had receded northward. All of them had moved at least 20 km, three quarters 50 km, a quarter 100 km, and 3% 300 km By January–February, marked locusts were being caught in the northern flood plains, That the initial movement took place southward was suggested by the absence of any marked individuals amongst locusts caught to west and north of the release sites, and by the high frequency of north winds from the ground to a height of at least 1,000 m with speeds greater than an insects' air speed. Likewise, movements by the North American **grasshopper** *Melanoplus mexicanus* in Montana and South Dakota during July (Munro & Saugstad 1938, Willis 1939) were consistent with the dominant south to south-east winds at the time of year.

Sometimes insects have characteristic properties that can be used as *natural markers.* On the night of 30 November 1967 at Trangie, New South Wales, there was an immigration of scattered **Australian plague locusts,** *Chortoicetes terminifera* (Clark 1969). Counts on experimental plots during the following day showed a considerable increase in numbers and their guts contained bits of plants not growing locally. Other evidence suggested windborne movement from a source about 100 km to the south (page 127; see also pages 146 and 180 for other discussion of this windborne migrant). Also in Australia, the sudden variations in degree of parasitisation of the **bushfly,** *Musca velustissima,* by nematode worms suggests that two or three waves of flies come to Canberra each spring, probably on spells of warm north winds from winter breeding areas further north (Hughes & Nicholas 1974; see also page 145). In West Africa, the **cotton stainer,** *Dysdercus voelkeri,* appears suddenly in large numbers after several months' absence in the dry season. At the beginning and ending of the rainy season there are two colour forms, orange and yellow respectively (Duviard 1977). To test the possibility that movement might explain these changes, four light traps were set up in Ivory Coast along a north-south line between about 5° and 9.5°N. Movement at the beginning of the rainy season was found to take place on the moist south to south-west monsoon winds in a zone a few hundred kilometres south of the ITCZ, The northern limit of populations during the monsoon (Fig. 73) is at about the 500 mm annual isohyet (equivalent to the northern limit of its malvaceous host plants), and the southern limit then is the northern limit of heavy monsoon rains (which drown the nymphs or induce fungal diseases). During the northern hemisphere winter this species is restricted to a narrow zone between the coast of the Gulf of Guinea and the ITCZ (which by then has moved to about 5-8°N).

Ramchandra Rao (1960) gives some examples of natural markers carried by

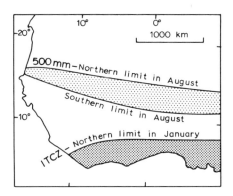

Fig. 73 – Seasonal variation in the northern limit of cotton stainer, *Dysdercus voelkeri*, in West Africa, due to movement of the ITCZ. (After Duviard 1977).

scattered **desert locusts.** *Schistocerca gregaria.* He found that the wings of old, overwintered locusts caught during spring in Pakistan sometimes had algae on them. Although these algae almost certainly could have developed only in the cool, damp weather of the Mekran coast, the marked locusts were caught up to 300 km to the north. Individual **African migratory locusts,** *Locusta migratoria migratorioides,* arriving from the north on the drying-out flood plains of the middle River Niger (page 147), were found by Davey (1959) to have the appearance of the generation that had developed in the surrounding dry lands during the previous rainy season. Moreover, all were young, immature adults that could not have been produced locally because the land had been flooded. There is substantial evidence that scattered individuals of both these locust species are taken downwind over distances of hundreds of kilometres (see pages 158 and 163 for *S. gregaria,* and pages 122 and 149 for *L. migratoria*). The sudden appearance of gregarious locusts (see page 169) among scattered individuals of any species is also indicative of invasion.

Sudden changes in other properties are also likely to be due to immigration from unknown sources. The following are some examples. During examination of sizes of some 6,000 *Spodoptera exempta* moths in East Africa during the 1973-74 armyworm season, Aidley & Lubega (1979) found that moths from a peak catch on the night of 16-17 February 1974 at Ilonga, Tanzania, had wing size ratio significantly different from all those caught there during January and the rest of February. Evidence for windborne movement by this species is presented elsewhere in this Chapter (pages 136, 150, 154 and 161). Again catches of the **large yellow underwing moth,** *Noctua pronuba,* in light traps at Rothamsted increased suddenly on the night of 1-2 August 1978 (Bowden *et al.* 1979). Until that night, females were largely immature, but subsequently there were some with large numbers of eggs. Moreover, two of the mature females caught on the night of 2-3 August were found to differ markedly in their composition of minor chemical elements as compared with earlier, immature individuals. The relative abundance of elements can be measured by X-ray fluorescence spectroscopy, but it is assumed not only that the pattern of relative abundance during feeding by the larval stage persists into the adult, but also that variations between individuals from a given source are small compared with differences between sources. Feeding during migration can alter the chemical composition, so adults being tested should have been migrating for no more than a day or two. This same method has been used with the **red turnip beetle**, *Entomoscelis americana* (Turnock *et al.* 1979).

The dominance of females in some species that increase in numbers suddenly is taken to indicate immigration. An example is the **6-spotted leafhopper**, *Macrosteles fascifrons*, which has been shown (page 134) to spread northwards each year across central USA on warm south winds (Meade & Peterson 1964). First arrivals in spring are mostly females, and females are dominant in summer even though populations reared in cages give approximately equal proportions of males and females. (See also pages 140 and 156 for further discussion of movement by this species.) In some species there are long- and short-winged forms, and laboratory experiments show that the long-winged forms are much more persistent fliers. Thus, the **brown planthopper**, *Nilaparvata lugens*, has been shown to be migrant to Japan (page 144). The first to arrive, in late June or early July, are long-winged whereas subsequent generations produced by local breeding are short-winged until late September or early October, when long-winged, emigrating forms appear again (Kusakabe 1979). (See also page 153 for further evidence in favour of long, windborne movements by this species.)

5.3 BACK-TRACKING, BUT NOT MARKING

An impressive and still increasing number of back-tracking studies have been made, suggesting that a wide range of species might be carried by the wind over distances up to thousands of kilometres. Some of these species are pests or carriers of disease, and they include both strong and weak fliers. Grasshoppers, dragonflies, butterflies and moths have been studied in this way, as well as beetles, mosquitoes, aphids and leafhoppers. Where windborne movements take place in no more than a few hours, they can be particularly easy to study because wind patterns may remain little changed, and a single windfield map is often sufficient for estimating a trajectory by eye. But longer movements, sometimes lasting a few days, require the careful construction of trajectories using a sequence of windfield maps (see page 74). We consider first some examples of windborne movements lasting no more than a few hours.

On 22 February 1970, out of more than 60 water traps in New South Wales only two caught the **cowpea aphid**, *Aphis craccivora*, and these were 80 km apart. There had been no local breeding, but back-tracking assuming take-off at 0700 h and flight until dusk led to likely sources on the coast of south-east Queensland, including an area of groundnuts, *Arachis hypogaea,* a known host of this species (Gutierez *et al.* 1974). Movement would have been on north-east winds on the north-west side of an anticyclone moving eastwards (see also pages 140 and 145 for other long movements by this species). A similar invasion by the **bird cherry aphid**, *Rhopalosiphum padi,* in south-east Denmark could be back-tracked on strong, warm south-east winds to Poland, where temperatures higher than in Denmark would have enabled winged forms to have developed earlier there (Thygeson 1968; see also page 141).

During World War 2, before the battle of El Alamein, Egypt, large numbers of the **mosquito**, *Anopheles pharoensis,* invaded a military camp in the early hours of 29 July 1942, about 50 km from the nearest possible breeding places and much further from any place where mosquitoes are likely to have bred in large numbers (Kirkpatrick 1957). They were mostly female (cf 124) and were probably brought from the Nile Delta on the east to north-east winds blowing at the time. A movement in the opposite direction involved the **greasy cutworm moth**, *Agrotis ipsilon.* On 9 May 1964, 200 were attracted to lights in Tel Aviv, Israel, where no local breeding was known (Odiyo 1975). On 7 May, a small but vigorous cyclone had moved north-eastward across the Nile Delta (Fig. 74), with warm south-west winds on its southern side that almost certainly brought moths to Israel on the night of 7–8 May from extensive infestations in

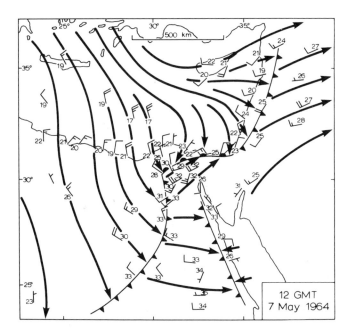

Fig. 74 – Weather map showing south-west winds over the Nile Delta that almost certainly brought greasy cutworm moths, *Agrotis ipsilon,* to Israel on the following night. See text.

cotton along the Nile valley above Cairo. The delay in capturing may be attributed to the moths resting and feeding after arrival in Israel and before flight was
resumed. For further discussion of this migrant, see pages 148 and 159.

In 1975, during studies of movements by Sahel grasshoppers of the middle
Niger area of Mali, a 3 cm pulsed radar showed a spectacular increase in aerial
density about 20 min after sunset, possibly due mostly to the **Senegalese grasshopper,** *Oedaleus senegalensis.* Back-tracking of this peak catch led to places
near the river Niger (Fig. 75) where surveys at the time showed that grasshoppers
were concentrated (Riley & Reynolds 1979). (See page 94 for further discussion of the use of radar.)

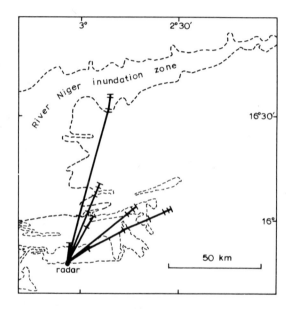

Fig. 75 – Back-tracks of grasshoppers (probably mostly *Oedaleus senegalensis*)
seen by radar in Mali on several nights during November 1975, leading back to
likely sources; It was assumed that take-off at source took place at the same time
as that of insects within radar range, and that ground speeds and tracks were the
same as those of insects within radar range. (After Riley & Reynolds 1979).

Massive movements of the **Australian plague locust,** *Chortoicetes terminifera,* have been associated with wind systems recognisable on windfield maps
(Clark 1969, Farrow 1977), and so have movements of the **Colorado beetle,**
Leptinotarsa decemlineata. Vast numbers of the **locust** came suddenly to light
traps at Trangie, New South Wales, on 30 November 1967, the first at 2125 h.
Flights during the night were also noted at other places along a 40 km north-
south line through Trangie. The most likely source was about 100 km to the
south, where streams of flying locusts had been seen during the three previous

days. If take-off on the 30th had been around dusk, giving a flight duration of about 3 h, downwind displacement in the southerly wind of 20-25 km/h would have been enough to bring the locusts to Trangie at the time they were seen. In fact, winds changed from west-north-west to south with the passing of a cold front across Trangie at about 1600 h (Fig. 76). The south wind had a speed of 5-10 km/h at a height of 1 m but could well have been 20-25 km/h at the height of locust flight, bearing in mind the likely strong wind shear near the ground at night (page 38). See also page 146 for this species.

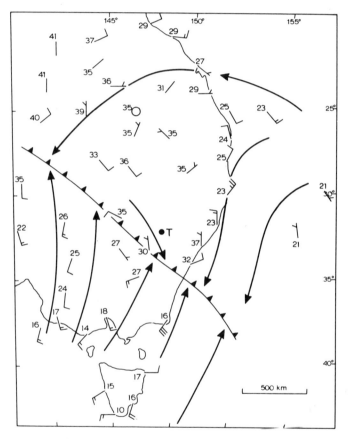

Fig. 76 – Weather map for eastern Australia, 1500 LT 30 November 1967, showing south and south-west winds behind a cold front that arrived at Trangie, New South Wales (T) late in the afternoon, and that brought an influx of scattered Australian plague locusts, *Chortoicetes terminifera*. See page 126.

The **Colorado beetle** is a potato pest well known for its progressive spread, first across North America and later across Europe. Its flights are thought to be short, and movement is often attributed to carriage in vegetables, but some idea of the distance that can be flown is given by invasions of the Channel Islands. In 1947, the first live beetles were seen in Jersey on 28 May, and in Guernsey on the next day. By 4 June, three more had been caught in Guernsey but nearly 400 in Jersey (Girard 1947, Dunn 1949). The 28th was the first day of a week of south-east winds and afternoon temperatures above 30°C widely over north-west Europe (Fig. 77). Because of the coincidence of dates it seems likely that at least some beetles flew the 50–100 km from France to the Islands, and some were indeed seen flying over the sea off the north-east coast of Jersey. Moreover, most catches were at the eastern end of that island. Other beetles, perhaps almost all those that set out from the coast of France, no doubt fell into the sea, for many dead ones, as well as some alive, were later washed up on the island beaches (Thomas & Wood 1980). A similar sudden arrival took place at the southern tip of Sweden on 5 June 1972, during a spell of hot south-east winds blowing from Poland (Gränsbo 1980).

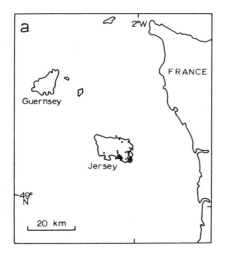 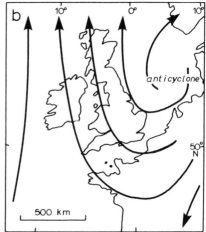

Fig. 77 – Distribution of Colorado beetles, *Leptinotarsa decemlineata*, reported on Jersey, 27 May to 6 June 1947. (After Dunn 1949). The windfield for 1800 GMT 28 May is representative of the hot east winds blowing from France to the Channel Islands that brought the beetles.

Cross-water movements are impressive when they involve flights of hundreds or thousands of kilometres, lasting up to several days. Movements from Australia to New Zealand are particularly clear. The **grain aphid**, *Macrosiphum miscanthi*, was found to be common on wheat throughout the Canterbury area of New Zealand in November–December 1967. It had never been found on wheat in New Zealand during the previous 11 years, despite regular field surveys. Infestations were in random patches, suggesting the settling of a few individuals here and there, acting as primary sources of spread. It seems that aphids had arrived suddenly: development rates suggest the period late October to early November. During the period 10 October–4 November there were only two spells when winds could have brought aphids to Canterbury from Australia; in both, the back-tracks led to near Melbourne (Close & Tomlinson 1975). Large numbers of aphids had probably been present on cereals throughout Victoria during the spring of 1967, and the winged forms would have been ready to fly in October–November as the crops matured. The two spells started on 27 October and 1 November, and the flights across the Tasman Sea on warm north-west winds (Fig. 78) would each have taken 2–2.5 days, with aphids arriving at Canterbury about the time that cold fronts were moving north, to be followed by changes

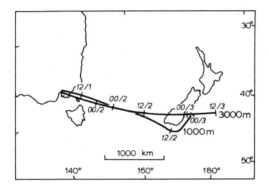

Fig. 78 – Back-track at heights of 1,000 and 3,000 m for grain aphids, *Macrosiphum miscanthi,* assumed to have been carried downwind and arriving in the Canterbury area of New Zealand at 00 GMT 3 November 1967, suggesting origins near Melbourne, Victoria, and an over-sea flight lasting 2–2.5 days. Ticks along the back-tracks show points reached at the stated times (GMT)/dates. (After Close & Tomlinson 1975).

to cool south winds. These movements may be compared with those of fungal spores to New Zealand (page 79, Fig. 40). Mention may be made here of two other insect species apparently crossing the Tasman Sea. Numbers of the **blue moon butterfly**, *Hypolimnas bolini nerina,* were seen at Nelson on 23 April 1971, persisting for a few weeks, but not surviving the winter. Back-tracks from 1200 h on that day, using wind reports at three different heights, suggest that the butterflies had come from Victoria or southern New South Wales on north-west winds (Tomlinson 1973; see Fig. 79). On 11 April 1977, two specimens of a **dragonfly**, *Pantala flavescens,* were seen on the north-west coast of the South Island of New Zealand. Windfield maps indicate a likely movement from about 30°S on the east coast of Australia, leaving on 8 April and arriving on the 10th (Corbet 1979). Supporting evidence for the long-distance flight ability of this species is provided by the six specimens caught on board ship south-west of Japan early on 26 June 1976, near a cold front separating northerly from south-easterly winds, but the source of these insects is unknown (Suzuki *et al.* 1977; see also page 101).

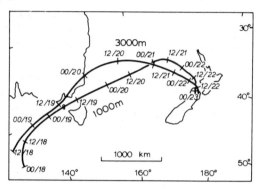

Fig. 79 – Back-tracks at heights of 1,000 and 3,000 m for blue moon butterflies, *Hypolimnas bolini nerina*, assumed to have been carried downwind and arriving at Nelson, New Zealand at 00 GMT 23 April 1971, suggesting origins in Victoria or southern New South Wales, and an over-sea flight lasting 3–4 days. For explanation of ticks, see Fig. 78. (After Tomlinson 1973).

Turning to north of the equator, cross-ocean movements have also been examined there. In Japan, increases in numbers of the **oriental armyworm moth,** *Mythimna separata,* during early summer are difficult to explain by local breeding but are consistent with movement in late July from China. An example is given by Oku & Koyama (1976), when there were strong south-west winds as a deep depression moved quickly from Kirin province to north Japan. Trap catches support the deduction of a movement at that time. (See also pages 119, 137, 144, and 152 for this species). In North America, the **monarch butterfly,** *Danaus plexippus,* is a well-known migrant (see page 95). Its flights between Canada, USA and Mexico are apparently within its boundary layer, but winds occasionally carry it away from its usual migration routes. For example, on 4 September 1970, some arrived in Bermuda, and back-tracks led to the coast of New England on the afternoon of 1 September – an over-water flight of about three days, first in north-west winds blowing off the coast, and later in north-easterlies across Bermuda on the east side of an anticyclone (Hurst 1971; see Fig. 51). Comparably long over-water flights have brought this butterfly across the Atlantic Ocean to Europe. During October 1968, starting on the 2nd, there were 60 reports from central and southern England of this rare visitor to Britain. Back-tracking led to sources in Canada if arrival had been about midnight of 28–29 September or the afternoon of 29 September, taking about two days for the crossing (Hurst 1969a). It was also found that all arrivals since 1933 could be back-tracked to the eastern seaboard of Canada or New England, taking 2–4 days for the crossings. Similar trans-Atlantic flights are able to account for the presence of the moth, **Stephens gem,** *Autographa biloba,* another rare visitor to Britain, but a native of North America. Individuals were seen on

19 July 1954 and on 1 October 1958. Assuming arrivals were on 17 July and 30 September, respectively, both could have come from eastern USA in 3–4 days on strong west to south-west winds (Hurst 1969b; see Fig. 80). The eastern Atlantic Ocean has also been crossed by the **small mottled willow moth**, *Spodoptera exigua*, flying from north-west Africa to England. Caterpillars of this moth, known variously as the **lesser armyworm**, **beet armyworm** and **pigweed army-worm**, are a widespread pest, feeding on the leaves of most plants. It is unlikely

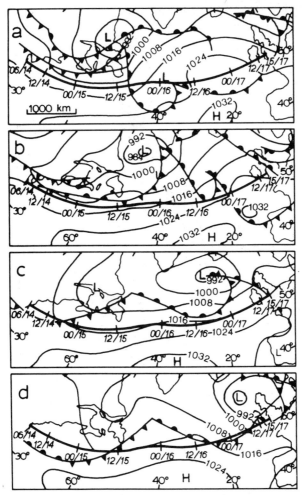

Fig. 80 – Sequence of 1200 GMT weather maps showing fronts, isobars (pressure in millibars) and a geostrophic back-track for the moth, Stephen's gem, *Autographa biloba*, (a native of North America but rare in Britain), assumed to have been carried downwind and arriving at Aberystwyth, Wales, at 1500 GMT 17 July 1954, a to d, 14 to 17 July. The back-track suggests an over-sea flight lasting about 3 days. For explanation of ticks see Fig. 78. (After Hurst 1969b).

that this species overwinters, as pupae, in Europe north of about 44°N, yet it
is seen in Britain almost every year. On 6 May 1962, the numbers in southern
England were more than had been seen for many years. Moth sightings and
catches by many amateur entomologists showed that arrival time was around
mid evening, and most came in from the sea at about 50°50′N 01°35′W. Such
timing and placing enabled a precise back-track to be built up, starting at 1800 h
on 6 May and going back to 1200 h on 4 May, before which time back-tracking
was difficult because winds were light near the middle of an anticyclone (Fig.
81 — this track is marked *A*). Assuming a moth air speed of 1 m/s, as well as a

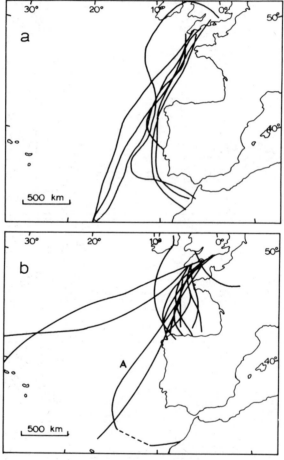

Fig. 81 — Back-tracks for various arrivals of small mottled willow moths, *Spodoptera exigua*, in southern England during the period 1947–67, showing likely
origins in Madeira and Morocco in February and March, but Spain or France in
May and June, although *A* is a back-track to Morocco for moths arriving on the
evening of 6 May 1962, assuming flight height decreasing for 1,200 m at source
to 30 m on the English coast. (After Hurst 1969c).

northward heading (as was suggested for a spring migration), trajectories showed that the most likely place and date for the source was Morocco on 2 May (Hurst 1969c). Thus there seems to have been a spell of unbroken flight over the sea lasting four days. It may well be that few of the moths setting out from Morocco reached England. Winds could have brought moths from north-west Spain on 5 May, or from north-west France on 6 May, but the numbers in England must have come from a source far greater than any likely to have been in Spain or France during May. Mild south-west winds, like those blowing on 6 May 1962, were found on all other notable moves by this moth to England from 1947 to 1963, with inferred sources early in the year in Morocco or Madeira, but from May onwards in north-west Spain. See also page 136 for another example of windborne movement by this species.

Other movements to Britain have come from the east. In 1976 there was an invasion of the **mourning cloak**, or **Camberwell beauty butterfly**, *Nymphalis antiopa*, a species known to overwinter sometimes in Britain, but never to breed This invasion was the greatest since 1872, and the butterflies were most numerous during the week 19-26 August, when the windfield was changing little: there was an anticyclone to the north-east of Scotland, leading to east winds from southern Scandinavia and Holland to England and Wales (Fig. 82). A back-track for the sighting on 22 August at Newark came from south-east Sweden in three days (Chalmers-Hunt 1977). The back-tracks from three earlier sightings came

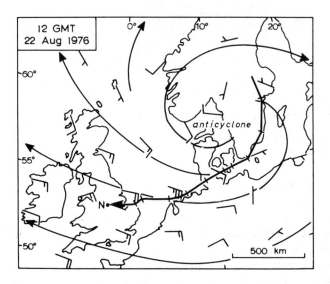

Fig. 82 — Surface windfield map for the time of arrival at Newark, England (N) of Camberwell beauty butterflies, *Nymphalis antiopa*, seen there at 1200 GMT 22 August 1976. Back-track in 6-h steps, showing the influence of a persistent anticyclone over Scandinavia. (After Chalmers-Hunt 1977).

from Denmark in three days. Similar tracks were taken by the **diamond-back moth**, *Plutella xylostella*, in 1958 (Shaw 1962). Caterpillars of this moth are a pest of cruciferous plants over most middle and tropical latitudes. At 1200 h on 4 July 1958 many moths were seen on a weather ship at 59°N 20°W. These moths were part of a great westward flight that reached north-east Britain on 28 June, later crossing mainland Scotland and Shetland to the ship, as well as to the Faroe Islands and Iceland (Fig. 83). Back-tracks led to western Russia, where weather in the ten days before would have caused moths to come from their pupae in large numbers. They were seen at the end of June in Finland, Sweden, Estonia, Norway and Denmark. See page 143 for another example of long-distance movement by this species.

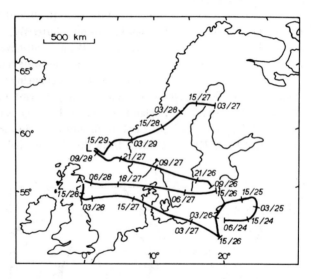

Fig. 83 – Back-tracks for various arrivals of diamond-back moths, *Plutella xylostella*, at Lerwick (L) and Aberdeen (A) during June 1958, suggesting sources in western Russia and flights lasting 2–3 days. For explanation of ticks see Fig 78. (After Shaw 1962).

Lest it be thought that such long-distance movements can take place only on the open sea, where settling to rest would almost certainly be fatal, mention should be made of back-tracked movements wholly over land. The **6-spotted leafhopper**, *Macrosteles fascifrons*, is the main carrier of the **aster yellows** mycoplasma, the cause of a disease in vegetable crops, barley and flax. Insects in Manitoba, Canada come on strong, warm south winds. The first specimens at Winnipeg in all years but one from 1954 to 1964 came in May. Back-tracks showed sources were in USA — the Dakotas, Nebraska and Kansas, all south of Manitoba (Nichiporick 1965; see Fig. 84). Eggs, but neither adults nor nymphs, can survive the winter of Canada and northern USA. Field surveys in USA

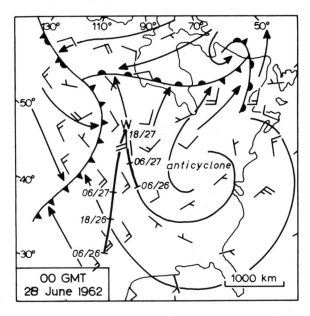

Fig. 84 – Surface windfield map for the time of arrival at Winnipeg, Manitoba (W) of 6-spotted leafhoppers, *Macrosteles fascifrons*, 1800 LT 28 June 1962. Back-tracks for surface (right) and 1,500 m (left) illustrate typical arrivals on southerly winds.

during the springs of 1953–58 demonstrated the northward movement of large numbers of insects to northern USA in southerly winds (Chiykowski & Chapman 1965). In Europe, the **gypsy moth**, *Lymantria dispar,* is a pest of many shade, fruit and woodland trees. Unlike many parts of the world, where this species is spread by young caterpillars being carried on the wind (pages 19 and 92), not by flight of adults, in Russia it has been seen on migration. For example, on the night of 25–26 July 1958 there was a huge flight near Moscow by an unusually pale-coloured form, distinct from the central European form, and on 28 and 29 July nine males were caught at six places as far as southern Finland. Windfield maps show that a cyclone moved slowly north-westward from the Black Sea from 26 to 28 July (Mikkola 1971). A warm front moved west across Moscow on the night of the 25-26th, and across Leningrad near midday on the 27th, and it became slow-moving along the west coast of Finland by midday on the 28th. The front was followed by warm south-east winds. A back-track from Tampere, where the first of the nine moths was caught near midday on 28 July, puts the Finnish moths passing north-east of Moscow, but one day later than the huge flight seen there (Fig. 85). This seemingly wrong timing might be due to flight having been fitful: it could have been mostly at night rather than continuous. The source of moths was most likely 200–300 km south-east of Moscow, among forests where about 300,000 ha of trees had had

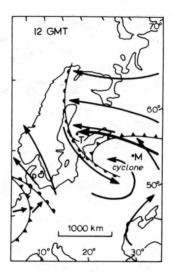

Fig. 85 – Weather map for 1200 GMT 28 July 1958, showing fronts, surface
streamlines and back-track for gypsy moths, *Lymantria dispar,* arriving at Tampere, Finland (T) at that time. The moths had probably passed Moscow (M)
during their flight from a source in western Russia. (After Mikkola 1971).

their leaves eaten. The 1,300 km flight to southern Finland is likely to have
taken 60–90 h, some of it on 3 or 4 nights. Again in Finland, the **small mottled
willow moth,** *Spodoptera exigua,* is an occasional visitor, but August 1964
witnessed a remarkable invasion (Mikkola & Salmensuu 1965). There were
captures at over 100 places, the first being in the south-east of the country on
the night of 4th–5th. By the night of 7–8 August moths were being caught in the
south-west, and also at Riga, in Latvia, and by 10–11 August they had reached
eastern Denmark. This spread fits with night-time flight on winds above the
planetary boundary layer, and back-tracking from Finland led to a source in
Soviet Central Asia, with departure on about 31 July. This was an area where
there had been extensive breeding during the summer. Appearance of moths in
Finland coincided with the westward passage of a warm front and the onset of
warm, south-east winds. Subsequent movement was mainly on north-east winds
around a slow-moving cyclone over western Russia.

    Rose (1979) and Blair *et al.* (1980) have shown by back-tracking that out-
breaks of the **African armyworm,** *Spodoptera exempta,* in Zimbabwe could
have been due to parent moths coming from Mozambique or Zambia. Back-
tracking of **fall armyworm moths,** *Spodoptera frugiperda,* that suddenly arrived
during the night of 3–4 September 1973 at Sault Ste. Marie, on the border
between USA and Canada, suggests an origin 1600 km away in northern Mississippi (Mitchell 1979). There is other evidence that this species is a windborne
migrant (for example, pages 155 and 158).

5.4 BACK-TRACKING FROM LATER DEVELOPMENTS

In addition to these well-documented case studies, there have been other back-trackings that have been based not on observed arrivals but on developments subsequent to inferred arrivals, whose timing can be estimated from development rates and compared with back-tracks to known or likely sources. Such developments include the appearance of a new generation, and of disease caused by organisms carried by the insects. In Japan, caterpillars of the **oriental army-worm moth,** *Mythimna separata,* can cause severe damage to grassland and cereals, especially rice. Sudden increases in numbers during the summer cannot be put down to local breeding and must be due to immigration of parent moths. From known caterpillar development rates, it is possible to calculate the approximate dates of arrival, and windfield maps show the days when moths could have come form their likely sources in China. For example, the extensive outbreak in 1971 can be related to moths arriving on 25 May from Honan province of China, where moths usually emerge from late May to early June (Oku *et al.* 1976). Similar movements have been deduced for 1969 (Oku & Koyama 1976) and for 1960 (Oku & Kobayashi 1977). In both 1970 and 1971, the sudden ending of infestations suggests invasion during a very short period – windfield maps again make it possible to narrow down the arrival dates to within a few days (Oku & Kobayashi 1974).

The timing of outbreaks of insect-borne **diseases** helps to pin-point arrival dates. From 6 to 8 August 1959 there were ten outbreaks of **malaria** at scattered settlements along the Israel coast north of Ashkelon, in an area considered to have been free of the disease for many years. Judged by incubation period of the disease, and by outdoor activities of the sufferers, 24 July is likely to have been the date when infected mosquitoes bit their victims (Garrett-Jones 1962). On the night of 23–24 July the weather was exceptional for the time of year – strong west to south-west winds with rain. Subsequent surveys showed the presence of the **mosquito,** *Anopheles pharoensis,* in 22 breeding places, but no other anopheline species; this species is rare in Israel and is therefore likely to have been the one involved. Moreover, the larvae were resistant to dieldrin, and resistance was already fully developed in parts of the Nile Delta, the nearest source. It is therefore likely that this malaria-carrying species came the approximately 300 km from the Delta (compare the movements of this same species within Egypt (page 125), and of the greasy cutworm moths from Egypt to Israel, discussed on page 125).

Also in the eastern Mediterranean, the onset in Cyprus during 1977 of an epidemic of **bluetongue**, primarily a disease of sheep, can be attributed almost without doubt to arrival of the virus vector – **midges,** mainly *Culicoides imicola.* Disease broke out within a week at both Kyrenia in the north and Famagusta in the south-east, parts of the island under different administrations and between which movement of sheep would have been most unlikely. First outbreaks were on 20 August. From known incubation periods within midges (7 days) and sheep

(6 to 10 days), the likely arrival dates for midges were 4–15 August. Winds over Cyprus during the first half of August were mostly from the west, as is usual for the time of year, but during two spells (5-6 and 11-14 August) they came from between north and east. The second spell fits best the reported range of first outbreaks, 20-25 August, and implies that arriving midges were already infective. Wind speeds indicate a flight duration of 5-20 h from southern Turkey, a likely source of midges, but only for flight at heights greater than about 500 m above sea level, for surface winds were light and variable (Sellers *et al.* 1979). In another outbreak of the disease, in southern Portugal during 1956, the inferred midge movement can be timed more precisely because winds able to carry midges from Morocco, the only likely source, blew only on the night of 21-22 June, when a cyclone moving north-westward over the sea between Morocco and Portugal led to a short spell of south-east winds – very unusual for the time of year (Fig. 86). Flight duration would have been about 10 h (Sellers *et al.* 1978). The same species of midge also carries the virus of **African horse sickness.** Studies of outbreaks in several countries have been made to examine the possibility of windborne movement of the vector (Sellers *et al.* 1977). For outbreaks in Cyprus in September 1960, Spain in October 1966 and the Cape

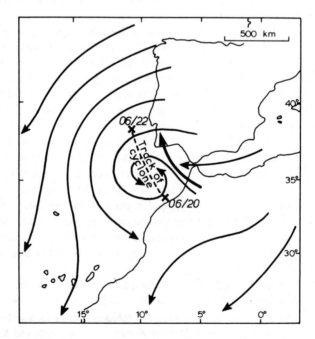

Fig. 86 – Surface streamlines at 0600 GMT 21 June 1956 showing the very unusual path of a cyclone for that time of year, and the associated track probably taken by midges, *Culicoides imicola,* bringing bluetongue from Morocco to sheep in Portugal. (After Sellers *et al.* 1978).

Verde Islands in January 1944, persistence from earlier outbreaks and movement of hosts were both most unlikely, so disease must have been due to movement of the vector. In 1960, the disease had appeared in south-western Iran during March, and then spread rapidly to Iraq by April, and to Syria and Turkey by May. This spread most likely took place in jumps — on spells of warm south-east winds between cooler north-westerlies, changes that are usual at the time of year. Although midges can fly at temperatures as low as 20°C, no spread seems to have occurred when afternoon temperatures in the south-easterlies were as high as 30°C, suggesting that flight took place at night — consistent with the already mentioned spread of bluetongue to Portugal. (After being taken over the sea, the midges may well continue flying by day if there is nowhere to land.) For outbreaks in September 1965 south of the Atlas Mountains in Algeria, the most likely source of infective midges was south of the Sahara. Winds there at that time of year are usually north-easterly, but August–September 1965 was a particularly disturbed period, with spells of south-west winds, two in September corresponding with the likely invasion period. In 1960, African horse sickness was diagnosed on 22 April at two places 320 km apart in north-western India, and the only known source was Pakistan, where the disease had been present at the end of March. Winds are usually very variable in direction at that time of year, and there had been two spells of west winds in mid April that could have brought midges from Pakistan to India.

The virus causing **ephemeral fever of cattle** is also likely spread by *Culicoides* midges. The disease has been found widely in the Old World tropics. In 1967-8 there was a big outbreak in Australia. It was first found near Darwin on 25 September 1967, after which it spread east and then south-east to reach north-western New South Wales and north-eastern South Australia by early February, and the east coast of Victoria by the end of February (Murray 1970; see Fig. 87). Such a wide and swift spread could not be put down to the moving

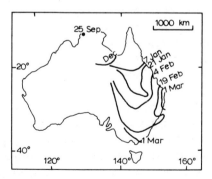

Fig. 87 — Spread of ephemeral fever of cattle from an initial outbreak near Darwin on 25 September 1967. This was almost certainly due to the windborne movement of *Culicoides* midges carrying the virus. (After Murray 1970).

of stock. Moreover, in the later stages there were enough data to show that live animals, in herds well apart from each other, were becoming ill along a front 500 km broad and moving at about 300 km a week. The most likely cause of this spread was the wind, for it could be linked with the onset of monsoon west winds over northern Australia, and with north-west winds on the eastern side of the slow-moving summer cyclones common over inland Queensland. Moreover, subtropical anticyclones moving slowly eastward over the Tasman Sea let northerly winds flow at times from Queensland into south-eastern Australia. Two such anticyclones gave strong, hot northerlies from 26 January to 2 February, and from 12 to 20 February. With an incubation period of about a week in cattle, these spells were found to fit very well with the spread of disease.

In the Central Province of Malawi, the virus causing **rosette in ground nut,** *Arachis hypogaea,* is not carried over from one year to the next in ground nut nor in wild hosts of the vector, *Aphis craccivora.* Hence the reappearance of disease in December–January is likely to be due to arrival of infective aphids from elsewhere – on the south winds of that time of year (Adams 1967). Similarly, the first adults of the **6-spotted leafhopper,** *Macrosteles fascifrons,* to reach Wisconsin in spring contain virus-carrying individuals, whereas locally bred leafhoppers need 5-6 weeks to become infective (Drake & Chapman 1965). This fact adds support to the deduction discussed on page 134 that this leaf-hopper migrates northward on southerly winds. In Sweden and Finland during late summer 1959 there was an unusual continental strain of **virus yellows in sugar beet.** Back-tracking showed that a flight from Germany within 24 h by the main vector, the **green peach aphid,** *Myzus persicae*, was possible on 6 July (Winktelius 1980).

5.5 ARRIVALS WITH WEATHER SYSTEMS

The above examples illustrate the long and growing list of insect species for which there are good grounds for believing that they are able to travel hundreds or thousands of kilometres on the wind. The list is impressive, yet it can be augmented substantially by many more examples of species that have been shown to arrive repeatedly with recognisable changes in wind direction. Thus, there have been many studies of insects arriving on spells of warm poleward-flowing winds in middle latitudes – southerlies in the northern hemisphere, and northerlies in the southern hemisphere. Other studies have associated arrivals with the monsoonal wind changes of low latitudes. Back-tracking has not been attempted because, for example, sources are unknown or they are too numerous for a choice to be made, but it seems very likely that movements are taking place that are similar to those already described where the evidence is more extensive and convincing.

Considering first the USA, all sudden arrivals of the **6-spotted leafhoppers** in Minnesota during spring were found by Meade & Peterson (1964) to accompany spells of south or south-west winds. The insects had probably come from southern states, where warmth had promoted earlier hatching. As shown above, such a source has been strongly suggested for the same species by back-tracking arrivals in Wisconsin (page 140) and in Manitoba (page 134). The **beet leafhopper,** *Circulifer tenellus,* is the only vector of **curly top disease** of sugar beet and other crops in western USA. It flies at dusk and dawn when temperatures are above about 15°C. In spring, leafhoppers fly to beet and other crops after overwintering on dense and widespread stands of weeds. For example, northward flight takes them from Arizona and New Mexico to Utah and Colorado. In the San Joaquim valley of California, the principal movements are in spring and autumn, between feeding sites in the valley during the summer, and in the western foothills during winter-spring. During evening flight, the dominant daytime northwest winds in the valley have been replaced by warm south-west winds coming across the hills and down the canyons, carrying the leafhoppers out into the valley. When large-scale winds become temporarily easterly, however, leafhoppers are taken westward across the hills, but they tend to stop flying when they meet the cool westerly sea breeze. It is these easterly winds of autumn, when the leafhoppers are leaving their summer feeding places, that carry the insects back to the western foothills because they tend to be warmer than the more usual north-westerlies (Lawson *et al.* 1951). Similar spring arrivals of aphids in South Dakota were found to be often associated with strong south winds (Kieckhefer *et al.* 1974). Daily trap catches at Brookings in March and April 1963-69 were sorted for **cereal aphids** *Schizaphis graminum, Macrosiphum avenae, Rhopalosiphum maidis* and *R. padi.* Of 427 trap days, 40 had cereal aphids, 98 had strong south winds, and 28 had both aphids and strong south winds. These results show there is a need to know the likely sources of insects if windborne movement is to be more convincingly demonstrated from association between arrivals and weather systems.

In 1975 a study was made of the possible long-distance movement of the devastating cotton pest, the **pink bollworm moth,** *Pectinophora gossypiella.* 170 traps baited with the synthetic pheromone gossyplure were set out in the desert area of southern California over an area of about 75,000 km^2 (Stern & Sevacherian 1978, Stern 1979). Moths were caught throughout the summer and winter, and no part of the desert was free of them. Short-range movements were shown to be normal, but the largest catches were in the Mojave desert coinciding with two tropical storms of wind and rain (on 7-11 and 16-17 September) moving north-west from southern to central California, suggesting movement from Arizona or New Mexico. (See pages 152 and 166 for further evidence of long-distance movement.) The **velvetbean caterpillar moth,** *Anticarsia gemmatalis,* does not overwinter in Mississippi yet moths were caught during the periods 29 March–3 April and 16–21 April 1978, three months before they are usually seen (Buschman *et al.* 1981). Weather maps showed that these arrivals corresponded to two spells of three or more days (1–6 and 14–17

April) with trajectories from southern Florida (the likely source), thus narrowing down the inferred arrival dates to 1-3 and 16-17 April. Further north, the **sunflower moth**, *Homoeosoma electellum,* does not overwinter in Saskatchewan, Canada, yet it is caught there occasionally in summer. Arrivals during 1978 and 1979 coincided with onset of short spells of southerly winds in late June and early July (Arthur & Bauer 1981). Warm southerly winds were also associated with an observation of the **dragonfly** *Anax junius* at Kitchener, Ontario on 4 April 1974 (Butler *et al.* 1975; see Fig. 88). On that day, afternoon temperatures of 20°C and winds of 40 km/h followed wintry weather for several weeks.

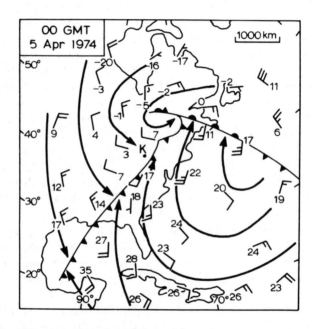

Fig. 88 – Weather map for 00 GMT 5 April 1974, showing fronts and surface streamlines and temperatures a few hours after the arrival of dragonflies, *Anax junius,* at Kitchener, Ontario (K) on warm south winds. By map time, cold west winds had returned to Kitchener behind the cold front.

The insects were sexually mature, thus ruling out the possibility of local emergence and implying movement from much further south. On this day there was widespread tornado damage in the USA, especially at Xenia, Ohio, from where papers were carried 300 km on the wind to Cleveland, Ohio. No further *A. junius* were seen at Kitchener until 14 May, when again there were 40 km/h southerly winds, and afternoon temperatures of 25°C (the first day exceeding 15°C for two weeks). An autumn sighting of another **dragonfly,** *Hemianax ephippiger,* in Iceland on 29 October 1971 corresponded with a spell of strong southerly winds on the 25–27th that could have brought the insects across the Atlantic Ocean from Morocco (Dumont 1976; see also page 100).

Turning now to *Europe,* the first arrivals of the **larch bud moth,** *Zeiraphera diniana,* around Lake Constanz, Switzerland, during 1973 were on 1 and 7 August, following spells of southerly winds bringing moths from the alpine forests of the Engadine to the south (Baltensweiler & von Salis 1975). Further north, the moth *Nycteola asiatica* is very rare. Reports in Finland, Sweden and Denmark are associated with spells of warm southerly winds from eastern and southern Europe, where it is common (Suomalainen & Mikkola 1967). Particularly impressive arrivals were on 3 October (southern Sweden) and 4 October 1962 (south-western Finland); and on 17 September (southern Sweden) and 21 September 1965 (southern Finland). Such warm winds can carry insects to arctic latitudes. On the afternoon of 25 June 1978, the **diamond-back moth,** *Plutella xylostella,* was caught at 78°N 15°E on Spitzbergen, following a severe south-easterly wind storm (Lokki *et al.* 1978; see Fig. 89). Weather maps show that warm southerly winds had spread north across Finland on the previous day (when large numbers of the moth had been seen in Finnish Lapland), reaching Spitzbergen on the 25th, with a 1200 GMT temperature of 13°C at Bear Island (75°N 19°E). These winds, on the eastern side of a deep cyclone over the Norwegian Sea, were not especially strong, so the storm in Spitzbergen was

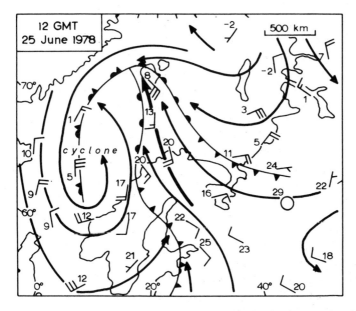

Fig. 89 – Weather map for 1200 GMT 25 June 1978, showing fronts and surface streamlines and temperatures at the time when diamond-back moths, *Plutella xylostella,* were arriving in Spitzbergen on south-east winds from Finland. See also Fig. 83.

probably a local one, almost certainly due to accelerating flow down the leeward side of the mountains (page 52 and Fig. 25). The movement was of the same kind as that more fully documented for this species by back-tracking from Scandinavia to Britain in 1958 (page 134). Again in Scandinavia, the incidence of **virus yellows of sugar beet** in Sweden by mid September increases with the frequency of winds with a southerly component during the summer — consistent with the movement of its main vector, the **green peach aphid**, *Myzus persicae*, from central and eastern Europe (Winktelius 1977; see also page 156).

The **oriental armyworm moth**, *Mythimna separata*, feeds on cereals and a wide range of other plants over southern and eastern Asia, the Pacific Islands, eastern Australia and New Zealand. In China, where it has been known as a pest for more than 2,000 years, the moths fly in spring from southern and eastern parts into the north-east (Lin *et al.* 1963, Li *et al.* 1964). Peak catches on ships in the Chili Gulf and the Yellow Sea during May–September 1960 were particularly associated with southerly winds (Hsia *et al.* 1963), the most spectacular being on 5 June 1960. Association of south-west winds with arrivals of this species in Japan has already been discussed (pages 130 and 137).

The **black**, or **greasy, cutworm moth**, *Agrotis ipsilon*, appears along with *M. separata* in China, supporting the evidence from elsewhere (pages 125 and 159) that this species, too, is a migrant (Wang 1980). The **brown planthopper**, *Nilaparvata lugens*, is another pest that moves northward over China each spring on spells of southerly winds. This species can overwinter only in the extreme south, where rice is present throughout the year (Cheng *et al.* 1979, Jiang *et al.* 1981). As seasonal temperatures rise, insects spread north to as far as about the Great Wall, the number of generations decreasing at higher latitudes. A system for forecasting movement in China started in 1979. This species does not overwinter in Japan, but comes to the paddy fields each June or July, sometimes as mass flights. Trap catches are greatest in south-western Kyushu, giving grounds for the view that the insects come from the south-west, almost certainly from China (Kisimoto 1976, 1979). Arrivals are mostly associated with depressions moving from central China to the Sea of Japan (Fig. 90), particularly near cold fronts and apparently only in those south-west winds ahead of the fronts that have come from likely sources. During field work to eradicate the **melon fly**, *Dacus cucurbitae*, from Kume Island (26°N 127°E), three million sterile flies were released each week during the period September 1975–August 1977, all marked with fluorescent dye. Pheromone traps put on four islets up to 50 km to the north-east were emptied twice a month from June to July 1977, and of 385 flies caught 15 were marked. The marked flies were caught only in May and June when south-west winds were most frequent, suggesting movement on seasonal monsoon winds and not under the influence of exceptional winds accompanying typhoons (Kawai 1978).

In Australia, caterpillars of the **southern armyworm moth**, *Persectania ewingii*, are a sporadic pest of cereals and pasture in the south-eastern and

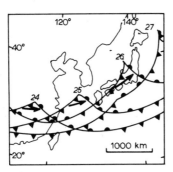

Fig. 90 – Changes in 0600 LT positions of fronts as a cyclone moved north-east from China across Japan, 24 to 27 June 1969, with a mass arrival of brown planthoppers, *Nilaparvata lugens,* on south-west winds represented by the arrows. (After Kisimoto 1976).

south-western parts of the continent. In Tasmania, it cannot overwinter (Helm 1975), and sudden outbreaks follow mass flights in spring (Martyn & Hudson 1953, Miller *et al.* 1963) in association with spells of warm north-west winds blowing from warmer source areas, possibly in New South Wales and South Australia (Drake *et al.* 1981). A mass movement on 29–30 September 1973 even led to arrivals in New Zealand. Light-trap catches of night-flying moths at Samford, Queensland were found by Persson (1976) to be strongly peaked on nights when cold fronts were passing and when the preceding north-west winds are likely to have come from sources furthest north. Dominant species included the **tobacco cutworm moth,** *Spodoptera litura* and the **tobacco looper moth,** *Plusia argentifera.* The Australian **bushfly,** *Musca velustissima,* is widespread, its grubs feeding on animal dung. It can be very troublesome, pestering man and his animals by seeking sweat and mucus (Hughes & Nicholas 1974). Southward spread across eastern Australia takes place on spells of warm northerly winds on the western sides of subtropical anticyclones moving eastward (Fig. 91), as with virus-carrying *Culicoides* (see page 139), the southern limit being where low temperatures stop flight. In the absence of a sea breeze, flies can be taken far out to sea. For example, on 3 and 4 January 1971, during a spell of offshore north-west winds between more persistent easterlies over Queensland, bushflies, moths and butterflies reached Heron Island, a small coral island about 60 km from the shore (Fletcher 1973). They had probably crossed more than 100 km of open sea, but many others perished for they were found dead in the sea even 50 km from the coast. The **cowpea aphid,** *Aphis craccivora,* feeds on a wide range of mostly leguminous plants in both tropical and middle latitudes. In New South Wales, it feeds in great numbers on grazing and weed legumes. In the spring of some years, clouds of this insect come to the middle coast, having been brought 300 to 500 km from large sources in the north-west. In 1951 the biggest clouds came on five separate days, the wind at a height of 1000 m

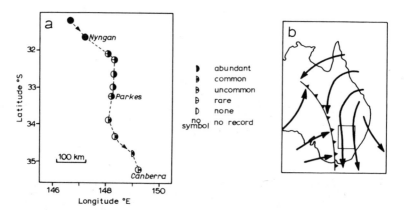

Fig. 91 – A southward movement of bushflies, *Musca velustissima*, in New
South Wales associated with north winds. (After Hughes & Nicholas 1974).
   a Changes in fly numbers from 11 to 12 September 1968 (left and right
      halves of symbols) at various places sampled along the route shown north-
      west from Canberra.
   b Streamlines at 0900 LT 12 September 1968 (the rectangle corresponds to
      the area of map a).

before four of them having been north-westerly for a day (and the same was true
of the fifth day if the cloud is taken as having come to the coast two days before
insects were first reported). Wind speeds were strong enough to need flight times
of less than a day. In 1948, many aphids came to the foreshore at Sydney in
the early afternoon of 10 October, at a time when a sea breeze was blowing. On
the day before, upper winds had been off-shore, and it seems likely that the
aphids had come from the north-west but had been taken out to sea before
being brought back to land on the sea breeze (Johnson, B. 1957). Studies of
immigration by scattered **Australian plague locusts,** *Chortoicetes terminifera,* at
Trangie, New South Wales, south-east of the main breeding area, during the
three breeding seasons of 1971–74, by means of field sampling four times a week
and light-trapping nightly, showed that of 28 occasions when more than 100
were caught in the light trap 23 were on nights with warm northerly winds,
associated either with southward extensions of tropical troughs of low pressure
or with cold fronts approaching from the south-west (Farrow 1979). The other
five occasions were on nights with southerly winds, emphasising the point that
this species migrates on warm, windy nights irrespective of wind direction, but
only if there has been sufficient rain during the hopper stage to provide good
pasture, probably because the abundant grasses then allow the build-up of enough
fat for flight (Symmons & McCulloch 1980). (A flight at Trangie in *southerly*
winds was discussed on page 126). The **spur-throated locust,** *Austacris guttulosa,*
also moves at night on warm north winds ahead of cold fronts advancing from
the south-west (Casimir & Edge 1979).

Comparably dramatic southward movements by scattered **African migratory locusts,** *Locusta migratoria migratorioides,* take place in north winds over *West Africa* behind cold fronts of the dry season. For example, daytime field surveys at Dioura, on the dry land around the middle Niger flood plain in Mali, revealed a marked fall in numbers of locusts between the afternoon of 25 and the morning of 27 November 1954. A violent dust-laden north wind set in during the late afternoon of the 25th, reaching Kara, on the flood plain 70 km to the south, during the evening, and led to peak catch in the light trap there on that evening (Davey 1959). Other sudden departures around the end of the year coincided with exceptional weather:

- strong north winds at night
- cloudy or dusty skies
- marked falls in day maximum temperature
- slight rise in night minimum temperature
- rising relative humidity by day and falling by night.

These weather changes are characteristic of the passage of a well-marked dry season cold front in West Africa, when light east winds north of the ITCZ are replaced by strong, cool north winds. Dew points first rise as the ITCZ approaches from the south, leading to a reversal of the usual fall in relative humidity by day. Inflow of dry air behind the cold front leads to a falling relative humidity at night. These abnormal trends would be most marked if the cold front passed through in the late afternoon or the evening. Similarly, Waloff (1963) associated changes in composition and density of populations of scattered **desert locusts,** *Schistocerca gregaria,* along the Red Sea coast of Ethiopia, with synoptic-scale changes in the windfield, when cold fronts moved south from Egypt.

Much longer, *open-water* movements, between Australia and New Zealand, have been strongly suggested by back-tracking, as discussed on page 128, and similar flights are almost certainly undertaken by a variety of moths and butterflies. Fox (1978) describes the recorded history of such flights and draws the following conlusions:

- conspicuous Australian species have been recorded in New Zealand since the earliest settlements
- several species are often reported at the same time
- many individuals of the same species are often reported at the same time
- the same species is often reported from widely separated places at the same time
- the species are often known to be migrants elsewhere
- recent arrivals are associated with north-west winds.

During the eight seasons 1968-9 to 1975-6, 36 Australian species of moths and butterflies were seen, sometimes associated with tropical cyclones. More direct evidence for open-water movements comes from sightings of the **greasy cutworm moth,** *Agrotis ipsilon,* a common garden and crop pest, from an oil rig off the west coast of New Zealand in January–February 1970 during spells of strong west winds (Fox 1970), and of the same species from a ship 500 km west of New Zealand in February 1957 (Common 1958).

Some of these examples of long-distance movements, for which there is more or less convincing evidence, are for first arrival in middle latitudes early in the season, on brief spells of poleward blowing winds warm enough for sustained flight. Later in the season, winds of all directions are more able to support sustained flight. The *C. terminifera* movements in Australia are examples (pages 122 and 126). Others can be chosen from North America. In early September 1953, the **mosquito** *Aedes vexans* spread to many parts of Illinois. Surveys between the 4th and 11th showed there were widespread infestations where no mosquitoes had been caught during several surveys in August. There had been no sites fit for breeding for two months and the most likely source was in Wisconsin, about 400 km to the north-west, where between 1 and 4 August there had been heavy rains that led to the formation of breeding sites. At the end of August and the beginning of September the weather was hot and dry, but a cold front moved south across Wisconsin on the 3rd, and across Illinois on the 4th, and the mosquitoes seem to have spread southward with the following northerly winds (Horsfall 1954). It has already been shown (page 118) that this species is taken downwind. The **forest tent caterpillar,** *Malacosoma disstria,* eats the leaves of many kinds of shade and forest trees in North America. Eggs overwinter on twigs, caterpillars hatch in spring, and moths come in clouds during summer. On 13 July 1964 many moths were seen in and around Calgary, Alberta, and a survey showed that they had come suddenly to large parts of southern Alberta. Their likely source was west of Edmonton, where many had been seen on 11 July. Weather maps showed that a southward-moving cold front had crossed the outbreak area in the middle of the afternoon of 12 July and had reached the far south of Alberta early on the 13th (Fig. 92). Again, displacement seems to have taken place in the northerly winds behind the front (Brown 1965).

Apart from these sudden arrivals in *middle latitudes,* there are others that occur with seasonal onset and retreat of the *monsoon* winds over Africa. In West Africa, field observations on changes in density of various species of **grasshoppers** in the sahel of Upper Volta strongly suggest that seasonal fluctuations are due to immigration and emigration, in association with changes in wind direction as the ITCZ passes over the country (Lecoq 1978). Adults arrive with the onset of north-east winds about October, lay their eggs in the soil and die. The eggs do not develop until there are sufficient rains about the following May or June, when the ITCZ has moved northward again and there are monsoon south-west winds over the country. After fledging, the new generation adults probably move away more or less to the north-east, to lay their eggs at a higher latitude. Such seasonal movements would allow the grasshoppers to take advan-

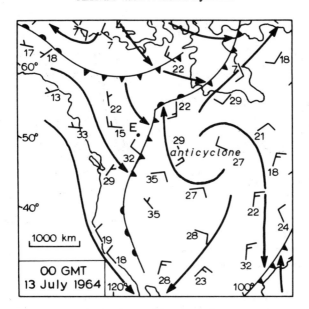

Fig. 92 – Weather map for 00 GMT 13 July 1964, showing fronts and surface streamlines and temperatures at the time when forest tent caterpillar moths, *Malacosoma disstria*, were moving south-east from near Edmonton, Alberta (E), behind a clod front.

tage of seasonally varying habitats, and they would help to explain upsurges following a sequence of drought years. At Kumasi, Ghana, Bowden (1973a) found large seasonal fluctuations of the **convolvulus hawkmoth,** *Agrius convolvuli,* a pest of sweet potato in Africa and south-east Asia, with peaks in February (a few weeks *before* the first rain maximum) and in October (a few weeks *after* the second rain maximum), as would be expected if the moths followed the movement of the ITCZ. Similar movements in the same part of West Africa by the **cotton strainer,** *Dysdercus voelkeri,* have already been discussed (page 122). In East Africa, hourly catches of grass-feeding Homoptera in suction traps on a 15 m mast at Quweiz village, about 80 km south of Khartoum, Sudan, in the cotton-growing Gezira area, showed increases when the wind changed from north to south, but decreases with reverse changes (Bowden & Gibbs 1973; see Fig. 93). These changes were attributed to north–south oscillations of the ITCZ, and to there being many more insects flying in the moist south winds to south of the ITCZ compared with those in the dry north winds on the north side. South of the equator, in Madagascar, scattered **Malagasy migratory locusts,** *Locusta migratoria capito,* have been shown by Lecoq (1975) to gather in the extreme south of the island during spells of north winds when the ITCZ moves well south in December–January, especially when there is a cyclone over the Mozambique Channel (see also page 147). Close to the equator, changes in

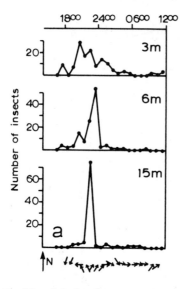
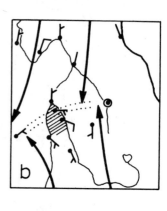

Fig. 93 – Relation between catches of grass-feeding Homoptera near Khartoum, Sudan, and passage of the ITCZ. (After Bowden & Gibbs 1973).
     a   Hourly winds at 15 m above the ground, and trap catches at three heights: 24–25 October.
     b   Surface winds and streamlines for 2000 LT 24 October, showing the ITCZ (double dotted line) and the extent of the cotton-growing Gezira (shaded).

light-trap catches of the **African armyworm moth,** *Spodoptera exempta,* at Muguga, near Nairobi, Kenya, have been associated with temporary wind shifts to the west from the dominant easterlies there (Haggis 1971). On the night of 9–10 March 1970, for example, clouds of many kinds of moths were seen. A light trap, modified to record hourly, caught 1,400 *S. exempta* in the seventh hour after sunset. The trap then failed due to overloading, but about another 7,000 moths would probably have been caught, judged by the 42,000 caught in a nearby standard light-trap (the highest nightly catch in seven years' trapping). There was a change of wind from east to west at 0050 h on the 10th. Weather maps showed that these westerlies were the leading part of a westerly airflow spreading from Uganda, reaching a line about north-south through Narok, Kenya and Dodoma, Tanzania, by 1400 h on the 9th. Douthwaite (1978), however, has re-examined the hourly catches and has shown that there are no significant differences in catches during easterly and westerly hours when wind speeds were no greater than 2.5 m/s (Fig. 94). Nor were there any significant differences in catches over 5-hour periods either centred on the hour with a windshift or without a shift, nor in hourly catches before, during or after rain. Only 4 of the 14 biggest catches were in a 5-hour period centred on a shift. Moreover, early in the armyworm season, catch size decreased progressively with increased windspeed. Douthwaite concludes that high catches on nights

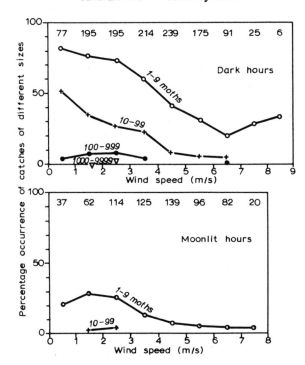

Fig. 94 – Relation between wind speed and occurrence of hourly catches of African armyworm moths, *Spodotera exempta,* of different sizes for both dark and moonlit hours at Muguga, Kenya, 1970-1. Numbers of hours in each speed range are shown. For example, at wind speed 2–3 m/s in dark hours, of the 195 hours, 25% had no catches, about 50% had 1–9 moths in each catch, about 15% had 10–99, about 5% had 100–999, and 5% 1,000–9,999. (After Douthwaite 1978).

with windshifts are to be attributed to improved trap catching efficiency due to light winds, not to the change in direction or any accompanying rain. This work draws attention to the difficulties in relating light-trap catch to the size of flying population without taking account of the effects of weather on both trap efficiency and insect flight. In a study of first arrivals of *S. exempta* moths at Nairobi, Brown *et al.* (1969) attributed them to a temporary strengthening of the north-east trade winds characteristic of the time of year (November–December). This species has been much studied in East Africa; a system of forecasting outbreaks has been in operation there since 1969 (see page 207).

There is circumstantial evidence (for example, page 137) that several insect-borne **virus diseases of animals** can be spread long distances from endemic areas within the tropics (between the seasonal limits of the ITCZ, where virus is present all the time but not seen as clinical disease because indiginous hosts are resistant) to epidemic areas outside, by means of the insect vectors being carried on spells of warm, poleward-blowing winds. Nearer the poles, the vectors cannot survive the winter cold, so virus is reintroduced either each year, or only occasionally at the extremes of the invasion area (Sellers 1980). The evidence is strongest for **Japanese encephalitis**, which reappears some time after both local emergence of the **mosquito** vector, *Culex tritaeniorhynchus*, and the arrival of migrant birds (the host, another source of virus), but soon after the onset of south-west winds, usually about mid June. This timing is very similar to that of the arrival of **brown planthopper**, *Nilaparvata lugens*; and like that insect, the mosquito has been caught at sea, far from land. Comparable movements occur in the southern hemisphere with **Murray Valley encephalitis**, carried by *Aedes vigilax* as far as Victoria, Australia; and in North America with **St Louis encephalitis** and **western equine encephalitis** (by *Culex tarsalis*), and with **eastern equine encephalitis** (by *Aedes vexans* and *A. sollicitans*), all three of which can reach as far as Canada. **Yellow fever** may also be carried this way by *Aedes* species.

5.6 SIGHTINGS FAR FROM SOURCES

When winged adults appear suddenly (and especially simultaneously at several places) and in such large numbers that they could not have been derived from the scanty local population (which also may be at a much earlier stage in its life cycle), then there must have been movement from elsewhere. Examples of species that have been seen this way are the **6-spotted leafhopper**, *Macrosteles fascifrons*, in the USA (Meade & Peterson 1964, Drake & Chapman 1965); the **brown planthopper**, *Nilaparvata lugens*, in China (Cheng *et al.* 1979); the **oriental armyworm moth**, *Mythimna separata*, in China (Lin *et al.* 1963); the **African armyworm moth**, *Spodoptera exempta*, in Uganda (Edroma 1977); and various moths and butterflies in New Zealand (Fox 1978). With all these species, there is also circumstantial evidence, discussed earlier, that their movements are downwind. It is not unreasonable to suggest, therefore, that arrivals of other species far from known sources are likely to have been due to windborne movements. In many cases, however, it is regrettable that neither back-tracking nor even association with particular wind systems has been attempted. For example, traps set up far from cotton at Eilat, Israel, caught **spiny cotton bollworm moths**, *Earias insulana*, a species widely spread over Africa and south-west Asia (Rivnay 1962), as well as **cotton leafworm moths**, *Spodoptera littoralis*, a species thought to be mainly non-migrant (Rivnay 1961). In contrast, *S. exigua* seen in Jerusalem on 13 May 1964 (Bytinski-Salz 1966) could well have come on a spell of unusual north-east winds from an unknown source. Similarly, when pheromone traps were set to catch **pink bollworm moths**, *Pectinophora gossypiella*, in a small cotton field in California at least 50 km from the nearest other cotton, moths of the new generation were caught from August onwards and caterpillars were seen afterwards (Bariola *et al.* 1973). Also in California, Kaae *et al.* (1977) showed that females caught in pheromone traps early and late in the growing season were distributed evenly over both cotton and adjacent maize, alfalfa and fallow, and that there was no variation of catch size with distance of non-cotton traps from cotton fields. This evidence is further indication of flight by this species from afar. It has been shown that movement of this species to southern California is likely to be on the wind (page 141); moreover, moths have been caught by aircraft at heights of 500 m and more above ground (Glick 1967). Similar experiments using traps baited with male **boll weevils**, *Anthonomus grandis*, at a cotton farm in Texas, 10 km and more from the nearest other cotton, caught many flying weevils; and traps in north-western Mexico caught weevils up to 60 km from cotton (Ridgway *et al.* 1971). Pheromone traps set up near Corpus Christi, Texas, where no cotton is grown, caught a few weevils in August, probably coming from Mexico (Pieters & Urban 1977). By using two fields in South Carolina, one with and the other without cotton, Roach & Ray (1972) found that numbers caught were similar in both fields and were unrelated to numbers in the cotton, thus indicating the presence of a migrating population. Moreover, weevils have been caught by aircraft at heights greater than 100 m above ground (Rummel *et al.* 1977), and it has already been mentioned that marked weevils have been shown to fly 50 km and more (page 121). This species is probably native to parts of Central America. It entered USA about 1892 near Brownsville, Texas, and subsequently spread about 100-200 km annually.

The **mountain pine beetle**, *Dendroctonus ponderosae*, is a periodic pest of most stands of lodgepole pine, *Pinus contorta*, in North America. Beetles emerge after a spell of warm, sunny weather, and fly to living trees of large diameter to bore into the bark and lay eggs in galleries. Bacteria and spores of fungi and yeasts are carried into the galleries and start to grow. **Bluestain** fungus, *Ceratocystis*, helps to kill the trees by restricting sap flow. The beetle has been found to cross timberless land 20-30 km wide (Safranyik 1978).

Sightings far out to sea are consistent with long-distance movements, possibly on the wind. Mention has already been made (page 148) of sightings over the Tasman Sea, between Australia and New Zealand, of the **greasy cutworm moth**, *Agrotis ipsilon*. Many **brown planthoppers**, *Nilaparvata lugens*, have been caught at ships in the East China Sea (Kisimoto 1976, Suzuki *et al.* 1977) and among the Phillipine Islands (Saxena & Justo 1980), and likewise the **oriental armyworm moth**, *Mythimna separata*, over the Yellow Sea and Gulf of Chili (Hsia *et al.* 1963). Catches over the East China Sea of aphid parasites (Mochida & Takada 1978) were associated with west winds blowing from China. Many **diamond-back moths**, *Plutella xylostella*, were seen over the Atlantic Ocean at 59°N 20°W on 4 July 1958 (Shaw 1962). There is strong evidence, already discussed above, that all four species are windborne migrants. The capture of **corn earworm moths**, *Heliothis zea*, at a light trap on an oil platform about 160 km off the Louisiana coast adds to the evidence (pages 155 and 166) that this species is a migrant (Sparks *et al.* 1975). Amongst other migrants caught on a similar oil rig have been moths of the **tobacco budworm**, *Heliothis virescens*, the **granulate cutworm**, *Agrotis subterranea*, the **velvetbean caterpillar**, *Anticarsia gemmatalis*, and the **fall armyworm**, *Spodoptera frugiperda*, as well as the pentatomid **bug** *Nezara viridula* (Baust *et al.* 1981). Night-flying **African armyworm moths**, *Spodoptera exempta*, a species for which there is much circumstantial evidence of downwind migration, were seen landing on a ship in the Gulf of Aden about 100 km from land on 10 June 1957 (Laird 1962). **Aphids** and **plant hoppers** often outnumber other species among catches over the Pacific Ocean hundreds of kilometres from land (Holzapfel & Perkins 1969; Guilmette *et al.* 1970).

Reappearance in an area previously cleared artificially provides evidence for migration. The most impressive example so far is the blood-feeding **black fly**, *Simulium damnosum*, the sole vector in West Africa of the parasitic worm that causes the disease, **river blindness**. Larvae of this fly require fast-flowing rivers for development; hence breeding sites are restricted. Despite very effective control of all known sites over a large area of West Africa, leading to a dramatic reduction in transmission of the disease, western and southern parts are reinvaded each year following the onset of monsoon winds. Invading flies are females that have already laid and taken a blood meal, and many carry the parasitic worms. From an extensive network of sampling points it has been found that reinvasion, as shown by sudden increases in numbers of flies coming to bite, are synchronous within a few days over areas hundreds of kilometres across (Le Berre *et al.* 1979, Garms *et al.* 1979). See also page 159 for this species. In the USA, the cold

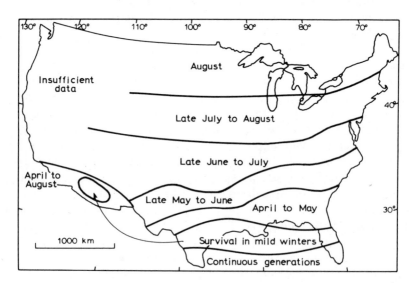

Fig. 95 – Average seasonal variation in distribution of infestations of fall army-worms, *Spodoptera frugiperda*. (After Young 1979).

winter of 1976–77 eliminated the **fall armyworm moth,** *Spodoptera frugiperda,* yet in the following summer there were widespread and severe outbreaks in the south-eastern states. Moths must therefore have reinvaded to account for subsequent populations, possibly from western Mexico, where in the preceding October there had been extensive rains from three hurricanes, or possibly from further south in Central America (Young 1979; see Fig. 95). Invasion from Mexico has been more clearly demostrated with the **tobacco budworm moth,** *Heliothis virescens.* Cotton around Brownsville, Texas, is planted in March, harvested in August and ploughed in by the end of that month, so overwintering in cotton is impossible. Overwintering is also unimportant in other crops and wild hosts, from which the peak emergence is in early April. In 1979, trap catches were largest in late March, too early and too great to have been derived from local sources; they must therefore have come from elsewhere (Raulston 1979). It has already been shown (page 119) from mark-and-recapture experiments that individuals of this species can move more than 100 km. A similar deduction can be made for the **corn earworm moth,** *H. zea,* which was found to arrive at Stoneville, Mississippi, one to three months before local emergence (Stadelbacher & Pfrimmer 1972).

5.7 SEASONAL REDISTRIBUTION

Seasonal disappearance and reappearance of flying adults at a given place can be due to seasonal changes in breeding, diapause, death or movement. More generally, varying parts of a population may move, diapause or build up and decline, depending on species, or within a species where there are variations in food supply, predators or other external pressures. Considering movement, its seasonality at a given place may depend more on the availability of populations elsewhere than on winds to carry them to that place. Thus, spring and early summer first appearances of the **6-spotted leafhopper,** *Macrosteles fascifrons* in the USA and of **brown planthopper,** *Nilaparvata lugens,* and **oriental armyworm moth,** *Mythimna separata,* in China and Japan, for example, are due to seasonal occurrences of warm winds that can bring insects from lower latitudes where they are abundant. By contrast, the late summer appearance of **pink bollworm moth,** *Pectinophora gossypiella,* and **boll weevil,** *Anthonomus grandis,* in the USA, and of **small mottled willow moth,** *Spodoptera exigua,* in spring in England, for example, are due to the seasonal development of a new generation ready to fly on spells of winds that have already been favourable for flight more than once during several months.

Studies of seasonal redistribution require field surveys and **biogeographical analysis** of the resulting records. In this way, the extent and timing of possible movements can be assessed. Of the biogeographical analyses that have been made, the most extensive and rewarding have been those on locusts, but before considering them we will look briefly at some other species.

With the **6-spotted leafhopper,** *Macrosteles fascifrons,* the evidence supporting windborne migration presented so far has been the sudden arrivals in Wisconsin and Canada associated with southerly winds before local populations
have hatched (page 141), the dominance of females (page 124), and the reintro-
duction of virus (page 134). Further support is provided by field surveys made
by car during the springs of 1953–58 which demonstrated that there was an
annual migration northward from western Arkansas, south-western Missouri and
nearby areas, to north-central USA and western Canada, sometimes over as
much as 1500 km (Chiykowski & Chapman 1965; see Fig. 84). Again in the
USA, the former summer reappearance of the **screw-worm fly,** *Cochliomyia
hominivorax,* was very likely on the wind (see page 120). The **green peach aphid,**
Myzus persicae, reappears each autumn in the Imperial Valley of south-eastern
California (Dickson & Laird 1967). A small network of sticky traps showed that
the aphids came across the mountains from coastal south California, probably
brought by the wind in much the same way that smog is well known to be
carried inland by coastal winds. In Britain, the seasonal redistribution of this
same species has been studied by means of a network of suction traps (Taylor,
L. R. 1977). By mapping 6-year average aerial density for each of the three
migration seasons – spring (from overwintering sites, mainly in southern England),
summer (mainly from the two principal crop hosts, sugar beet and potato), and
autumn (from a wide range of late summer hosts over the whole of England) –
the importance of spring migration from the south was demonstrated (Fig. 96).
It supported an earlier suggestion that such migrations would account for the
observation that the fraction of sugar beet infected by **yellows** in the summer

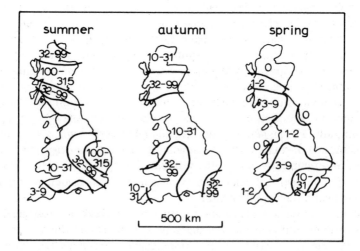

Fig. 96 – Seasonal variation in mean numbers of green peach aphids, *Myzus
persicae,* caught in suction traps 1970–5. (After Taylor 1977).

is strongly related to winter temperatures in *southern* England, not locally, even though crops may be up to 250 km from there. The analysis shows how this species effectively scans its environment and exploits available hosts without being able to control its flight direction most of the time. (See also pages 140 and 144 for movement by this species). The **damson-hop aphid,** *Phorodon humuli,* has also been studied with the same trap network (Taylor *et al.* 1979). This species overwinters as eggs on certain *Prunus* species. Eggs hatch in early spring and, after several viviparous generations, winged adults appear and fly in early summer to hops, *Humulus lupulus.* In early autumn, after more vivparous generations, winged adults fly to *Prunus,* where egg-laying forms are produced. Sources in England from this autumn migration are very restricted to the hop gardens of Kent and Herefordshire. Using seven years of daily trap records, the median distance travelled was found to be 15-20 km, and only about 5% moved more than 100 km (Fig. 97). At distances of several hundred kilometres, densities derived from these hop gardens may be comparable with those derived from scattered wild hop hosts. There was no evidence that spread was greater in some directions than in others — presumably reflecting the day-to-day variability of winds. See also pages 118 and 166 for movement by this species.

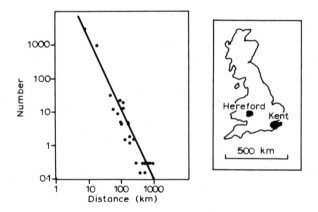

Fig. 97 — Variation of aerial density of hop aphids, *Phorodon humuli,* with distance of sampling point from the nearest principal source, either Kent or Hereford. (After Taylor *et al.* 1979).

In Asia, the **rice leaf roller moth,** *Cnaphalocrocis medinalis,* is a widespread pest in southern and south-eastern parts, and it has become serious in China since 1965. Neither caterpillars nor pupae can overwinter north of about 30°N, and at places further north it reinvades each year from the south (Chang *et al.* 1980). In Japan, appearances are associated with arrivals of the **brown planthopper,** *Nilaparvata lugens,* providing further circumstantial evidence that the moth is a windborne migrant (Miyahara *et al.* 1981). In the USA, several moth species overwinter only in the warmest, extreme southern, parts around the Gulf of Mexico, and they apparently spread north during the spring, presumably on warm south winds in much the same way as *Mythimna separata* in China (page 144) and Japan (page 130), and *Spodoptera exigua* into England (page 131). These species include the **beet armyworm moth**, *Spodoptera exigua* and **fall armyworm moth**, *Spodoptera frugiperda* (Mitchell 1979), as well as the **cabbage looper moth**, *Trichoplusia ni* (Chalfant *et al.* 1974) and the **soybean looper moth**, *Pseudoplusia includens* (Mitchell *et al.* 1975).

In Australia, it has been seen that there is evidence to show that the **bushfly,** *Musca velustissima,* spreads southward on warm north winds, to reappear each spring in, for example, the Canberra area. Fortnightly spring surveys for 1,500 km north-west from Canberra confirmed the progressive southward spread of flies (Hughes 1970).

It is sometimes claimed that the poleward spread of species such as these over middle latitudes in spring is followed by a return equatorward only in autumn, or even by a failure to return. But neither is likely, because equatorward movement can take place whenever there is a northerly wind component and a temperature above the threshold for flight. Temperatures in middle latitudes generally decrease polewards, so flight will be possible only on the equatorward side of the flight threshold isotherm (FTI). At any given time, the FTI will be a wavy line lying more or less west-east, although made more complex near mountains and coasts. It will move seasonally: poleward in spring and equatorward in autumn, but also with oscillations of a few days due to temporary weather systems. Hence the area of possible flight extends poleward in spring and retreats equatorward in autumn. Poleward extension usually takes place in poleward-blowing winds. Close to the FTI, both insects and FTI can be taken poleward together, leading to occasions, such as have been described, of new arrivals with warm winds. Insects will sometimes end up poleward of the FTI if it moves back across them when they are not flying – especially when they are in the egg or juvenile stages. Flight then being impossible, the insects may die or go into diapause. Such events are most likely to happen when the FTI is moving equatorward fastest, in autumn.

Biogeographical analysis of many years' field reports of solitary-living **desert locusts,** *Schistocerca gregaria,* has clearly shown a seasonal redistribution of whole populations over vast areas, similar to that of swarms (page 170) although more restricted in range. For example, in the deserts of north-west India and Pakistan, populations are sparse and breeding rare from December to June, whereas in western Pakistan populations are larger and breeding is common in those months (Ramchandra Rao 1960). The reverse is true from July to November. These differences are due to the seasonal incidence of rain, and movement on the seasonal winds. The same is true of the **African armyworm,** *Spodoptera exempta,* the caterpillar of a night-flying moth, and so-called because it tends to crawl in huge numbers, sometimes at densities greater than $1,000/m^2$.

It feeds on wild grasses as well as cereals and pasture, in eastern and southern Africa, and can cause as much harm as the locust. Outbreaks vary in size from year to year. At a given place there are large seasonal changes in the numbers of caterpillars, and outbreaks have a strong tendency to follow one another in place and time, as was pointed out in 1943 by Faure. Thus, the first outbreaks in east Africa are often in Tanzania at the end of the year, and later outbreaks are farther and farther north, reaching Ethiopia by the middle of the year (Brown et al. 1969). The suddenness of these seasonal outbreaks, after months with none in the vicinity, takes farmers unawares and gives grounds for thinking they came from a sudden inrush of parent moths. This view is supported by the fact that there seem to be no differences in enzyme systems from 17 specimens caught up to 2,000 km apart in east Africa, indicative of continual mixing of populations over the area sampled (Den Boer 1978). As yet, moths have not been tracked far downwind, but such a movement seems very likely, because they have been seen to move downwind, both by eye (page 161) and by radar (page 196). Downwind displacement would take moths towards and into the ITCZ, so that most of the insects, whether caterpillars or moths, would be near the ITCZ, where rains fall and grasses grow. Moreover, the following generations would move with the ITCZ as has been shown by catches from a network of moth traps over east Africa. Huge numbers of moths carried downwind can lead to eggs being laid in very close proximity. The subsequent dense masses of caterpillars pupate in the soil after about two weeks and the moths emerge and leave at night, often unseen. As the part played by weather in the timing of outbreaks became better understood, a forecasting service was initiated in 1969 (see page 207). By contrast, spread of the related species, the **cotton leafworm moth**, *Spodoptera littoralis*, monitored by a network of pheromone traps in Cyprus, was found   by Campion et al. (1977) to be unaffected by the wind. Populations spread slowly, and flight was presumably within the insect's boundary layer. (But see also page 117).

Another noctuid, the **greasy cutworm moth**, *Agrotis ipsilon*, occurs widely in middle and tropical latitudes, and its caterpillars attack a wide range of wild and crop plants, characteristically cutting off seedlings at the base of the stem. Sudden and seasonal disappearances have long suggested that this species is a migrant, and mapping its distribution month by month has shown that the latitudinal range of the adult is greater than that of the caterpillars in both northern and southern hemispheres (Odiyo 1975). This species is probably unable to overwinter at the poleward limits of its distribution, and it seems to spread poleward each year (see pages 125, 148 and 153).

Again in Africa, we have seen that the **black fly**, *Simulium damnosum*, the vector of the parasitic worm causing **river blindness**, is a migrant that invades areas cleared by treating the breeding sites with insecticide. Study of winds associated with invasions shows that they occur mostly in the monsoon south-westerlies, to south of the ITCZ. But because of the day-to-day variability of

winds, particularly due to thunderstorms, it is not wise to assume that movement is simply from south-west to north-east. Moreover, because the delay
between arriving and biting is unknown, trajectories from particular sightings
cannot be calculated (Magor & Rosenberg 1980). Experimental control of
breeding sites of this species has led to decreases in numbers of flies caught
downwind. It has also demonstrated that flies arriving at a particular place come
from more than one source, as would be expected for any windborne species
because of day-to-day variability of the windfield (Walsh *et al.* 1981).

5.8 INDIVIDUAL FLIGHT IN RELATION TO THE WIND

Insects of several species, after flying up through their boundary layer, have been
seen to turn and head downwind. Newly-emerged **mosquitoes,** *Aedes cataphylla,*
at Edmonton, Alberta, after taking off into the wind, were seen to drift backwards before quickly turning downwind and climbing at about 40° to the
horizontal until reaching a height of 4 m, after which they flew straight and level
in a wind of about 1.5 m/s at 2 m (Klassen & Hocking 1964). On another
occasion the level flight was at about 8 m in a 3 m/s wind at 2 m. Downwind
flight soon after take-off has also been demonstrated for the **European elm bark
beetle,** *Scolytus multistriatus* (Meyer & Norris 1973). Sticky traps were set up in
two hemispheric grids (radii 1 m and 3 m) over a source of about 1,000 laboratory-
reared beetles at Arlington, Wisconsin, It was found that both the horizontal and
vertical angles of the cone of flight tended to decrease as the wind speed increased,
showing that the beetles had progressively less control over their flight direction
as the wind strengthened. The **pine weevil,** *Hylobius abietis,* was similarly
found to turn after take-off from the centre of a circular clearing in a forest
(Fig. 98), and then fly persistently downwind (Solbrek 1980). Grubs of this
weevil feed under the root bark of freshly-killed pine or spruce trees. This food
source is limited and adults migrate in spring or early summer. Similar flight

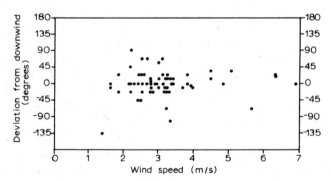

Fig. 98 – Directions of flight of pine weevils, *Hylobius abietis,* when 20 m from
a release point, in relation to wind speed and direction. (After Solbreck 1980).

was seen in the **green lacewing**, *Chrysoperla carnea* (Duelli 1980). Laboratory-reared insects were released upwind of alfalfa sprayed with an artificial food like their natural food, honeydew of homopterous insects. Lacewings, one or two nights old, flew straight downwind at heights above 3 m and were not attracted to food (Fig. 99). Older specimens, including some well-fed wild individuals of unknown age caught the previous night, also first flew downwind, mostly below 3 m, but after passing food they turned upwind and were attracted to the food. Laboratory-reared **mahogany shoot borer moths**, *Hypsopyla grandella,* the caterpillars of which cause stunted growth of Spanish cedar, *Cedrela,* and mahogany, *Swietenia*, were released within a plantation in Costa Rica (Holsten & Gara 1975). In winds greater than 0.5 m/s the moths first flew upwind, but quickly turned downwind.

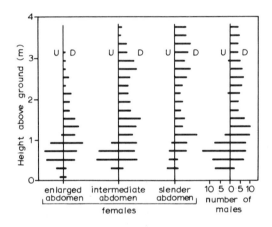

Fig. 99 – Height and direction (U = upwind, D = downwind) of flight of green lacewings, *Chrysoperla carnea,* in the centre of an alfalfa field at Kerman, California, during a 31-day period, showing increasing tendency of older females to fly upwind close to the ground. (After Duelli 1980).

In a field study of the **African armyworm moth**, *Spodoptera exempta,* near Nairobi, Kenya, it was found that night flights were mostly downwind (Brown & Swaine 1966). In a later field study, moths were seen to fly soon after sunset from daytime shelters under grass tufts, dung pats and tree bark, to the nearest tall trees, where at first they settled but later they milled about before flying off, *heading* downwind at a height of about 10 m (Rose & Dewhurst 1979). Nearly all moths had gone within about half an hour. By using an upward-pointing light beam, a peak in numbers passing overhead was found to last about an hour. Another peak, centred about 2200-2300 h, was almost certainly due to moths recently emerged from the soil. Earlier observations at nearby Muguga, also using a light beam, showed that insects in general (probably mostly moths) at this time of night were flying mostly at heights of 4-8 m above the ground,

90% of them downwind, to within one point of a 16-point compass (Brown 1970; see Fig. 100). Another moth, *Loxostege sticticalis*, has been found to fly downwind both by day and by night (Melnichenko 1936). The **silver-Y moth,** *Plusia gamma,* flies downwind at night but not by day, when it behaves like a butterfly, by keeping to its boundary layer (Larsen 1949, Taylor *et al.* 1973).

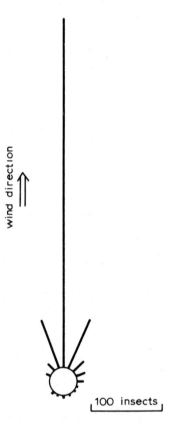

Fig. 100 – Numbers of insects (probably mostly moths) seen flying across a vertical light beam in 16 directions relative to the wind, on 8 nights at Muguga, Kenya. (After Brown 1970).

Solitary **locusts** fly by night and are caught in light traps, as are grasshopppers, contrasting with *swarms*, which fly by day (page 170). Indeed, only such night flying can account for the observed day-to-day changes in locust numbers. By using upward-pointing light beams in Arabia (Roffey 1963) and in Ethiopia (Waloff 1963), it was found that night-flying scattered **desert locusts,** *Schisto-cerca gregaria,* not only moved but also actively *headed* downwind in winds stronger than 3–4 m/s, about their air speed (Fig. 101). In lighter winds, movement was much more variable, as would be expected for flight in the locusts' boundary layer (page 93).

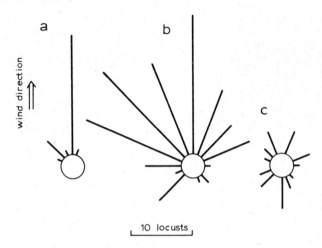

Fig. 101. – Numbers of scattered desert locusts, *Schistocerca gregaria*, seen flying across a vertical light beam in 16 directions relative to the wind in Saudi Arabia: a > 4 m/s; b < 2.5 m/s and locusts flying above 10 m; c < 2.5 m/s and locusts flying below 10 m. (After Roffey 1963).

In addition to these *eye* observations of downwind heading above the insect boundary layer, **radar** has been used increasingly to extend the range of sight. X-band marine pulsed radar (wavelengths about 3 cm) have been modified to detect individual insects at distances out to several kilometres, the distances increasing with insect size (Schaefer 1976, Riley 1978). By filming the radar display, the ground speed and track of an individual can be measured and, by vector subtraction of the wind (measured by, for example, radar-tracked balloon), the air speed and heading can be deduced, allowance being made for errors due to geometrical distortion caused by trying to represent, on a flat surface, echoes from the conical surface swept out by the radar beam. Identification of species is still a problem (Riley 1979) but is to some extent helped by amplitude modulation of the radar echo due to wing-beating and breathing, because wing-beat frequency is a guide to insect size. Schaefer (1970, 1972a, 1972b) used such a radar in Niger, West Africa, and showed that scattered **desert locusts,** *Schistocerca gregaria,* flying at night above their boundary layer, were almost always heading downwind. With average ground speeds of 30 km/h, it is possible that individuals could travel several hundred kilometres in a night. Heading, which can also be measured because side-viewing gives a stronger echo than head-viewing (Fig. 102), was remarkably uniform amongst many individuals, whether in moonlight or starlight, but the mechanism by which the locusts can detect wind direction is unknown; it may be by sight, even in such weak light. Schaefer

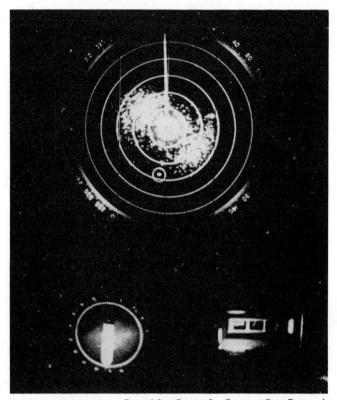

Fig. 102 – Radar screen showing uniform distribution of mixed insects but a degree of common heading that leads to a non-uniform display; there are more echoes in directions with side-on aspects, giving a 'bow-tie' effect. Kara, Mali, 1946 LT 3 November 1973. Elevation 46°; range rings 450 m apart. A radar-tracked balloon is circled. (After Riley 1975).

(1970) has suggested that the flying locust may be able to detect some structure in the wind turbulence that could indicate direction. In Mali, radar studies of flight by grasshoppers over the middle Niger flood plain showed that insects (most probably the **grasshopper**, *Aiolopus simulatrix*) flew either downwind or crosswind (Riley 1975), but Sudanese grasshoppers, mostly the same species, headed south-south-west in all but strongly opposed winds (Schaefer 1976). In Canada, night-flying **eastern spruce budworm moths**, *Choristoneura fumiferana*, were seen to be heading downwind (Greenbank *et al.* 1980), and so were **southern armyworm moths**, *Persectania ewingii*, in Tasmania, Australia (Drake *et al.* 1981).

These observations of flight turning downwind above the boundary layer support the optomotor hypothesis of navigation proposed by Kennedy (1939) in which the direction of heading is assumed to be controlled by the apparent motion of images of the background (see page 111). It seems that the individual insect has a *preferred air speed* (which may be felt by hairs on its face), and a *preferred angular rate of apparent movement of the ground* (from front to back, and seen by its eyes). If

V = insect's air speed

W = wind speed (positive if in the same direction as V – that is, for downwind heading)

ω = angular rate of apparent movement of the ground

z = height above visual pattern (for example, ground or vegetation),

then
$$V+W = \omega z$$

for flight up or down wind (Fig. 103). From this it follows that if V and ω are constant (that is, preferred values for a given insect), then z will increase linearly

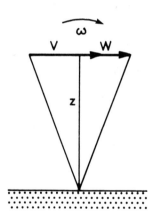

Fig. 103 – Angular rate of apparent movement of the ground, ω, for an insect flying at air speed V in a wind speed W at height z.

with W; that is, for *downwind* flight, the stronger is the wind the higher the insect flies to keep ω constant. This may at times lead to higher flight in stronger winds – for example, the mosquitoes studied by Klossen & Hocking (1964) (see page 160). Apart from this apparent control on height of flight, it is an advantage for long-distance, downwind movement to be high above the ground because wind speed usually increases with height (page 38). Mention has already been made that many species have been caught flying many tens or even hundreds of metres above the ground. The aircraft trapping records of Glick are among the earliest and most extensive. For example, the **pink bollworm moth**, *Pectinophora gossypiella*, the **fall armyworm moth**, *Spodoptera frugiperda* and the **cotton leafworm moth**, *Alabama argillacea*, among others have been caught over 50 m above the ground (Glick 1965, 1967), and both the **boll weevil**, *Anthonomus grandis*, and the **damson-hop aphid**, *Phorodon humuli*, at over 100 m (Rummel *et al.* 1977, Taimr *et al.* 1978). Using black-light traps on a 300 m TV tower near Pelham, Georgia, in mixed farming country, Callahan *et al.* (1972) caught **corn earworm moths**, *Heliothis zea*, at all heights, suggesting movement from maize-growing areas. Such observations lend support to the suggestion, more strongly indicated by other evidence presented earlier, that these species are windborne migrants. If there is a **low-level jet** (page 40), with its strong vertical wind shear, ω may not decrease with height until near the level of strongest wind, above which it will decrease rapidly. Flying insects with a preferred ω less than that produced by winds below the level of strongest wind may have to climb to near or even above the maximum before they can maintain level flight. Thus the optomotor hypothesis would explain the occurrence of elevated streams of insects at night shown up on radar screens as echo layers (page 191).

5.9 OTHER EVIDENCE FOR LONG-DISTANCE FLIGHT

Flight *duration* is another indicator of possible long-distance windborne movements by small, slow-flying species when they fly persistently above their boundary layers. Laboratory experiments with flight mills have shown that individuals of many species can fly for several hours, intermittently if not continuously. Examples are: mosquitoes, fruit flies, black flies, aphids, leafhoppers and planthoppers. Stronger fliers, such as beetles, moths and grasshoppers, have flown for 5–10 h, and even more then 20 h, and for longer still when fed between flights. For example, laboratory experiments by Gatehouse & Hackett (1980) with the **African armyworm moth**, *Spodoptera exempta*, supported not only the flight durations inferred from seasonal redistribution and windfields but also field observations of the timing of flight.

5.10 SUMMARY

This Chapter has presented a great variety of circumstantial evidence suggesting that downwind movement is common among many kinds of insects. The following species, perhaps among the best documented, are taken from this section and they illustrate the variability of evidence.

	Marking and back-tracking	Marking but not back-tracking	Back-tracking but not marking	Back-tracking from later developments	Arrivals with weather systems	Sightings far from sources	Seasonal redistribution	Individual flight in relation to wind	High-level flight
moths									
Agrotis ipsilon		125			148	153	159		
Heliothis virescens		119				154 155			
H. zea						153 155			166
Mythimna separata		119	190	197	144	152 153	155		
Pectinophora gossypiella					141	152	155		166
Spodoptera exempta		123	136		150 151	152 154	158	116	
S. exigua			131 136			152	155 158		
S. frugiperda			136			154 155	158		166
S. littoralis	117					152			
locusts									
Chortoicetes terminifera		122	126		146				
Locusta migratoria		122 123			147 149				
Schistocerca gregaria		123			147		158	162 163	
aphids									
Myzus persicae				140	144		156		
Phorodon humuli	118						157		166
planthopper									
Nilaparvata lugens		119 124			144	152 153	155		
leafhopper									
Macrosteles fascifrons			134	140		152	155 156		
weevil									
Anthonomus grandis		121				153	155		166

Many windborne movements appear to be purposeful, and they are sustained over hours or days, and sometimes weeks. They cannot be mere accidents, for each individual could often be able to avoid downwind movement by descending to fly in its boundary layer. With most species, however, there is a need for evidence that is direct, not circumstantial. Mark-and-recapture experiments are the most convincing, when combined with back-tracking in relation to the wind. If it is thought that a given species might be windborne, the literature should be searched for as many kinds of evidence as possible from among those described in this section. If such evidence is insufficient, field and laboratory experiments may be needed to supplement it.

CHAPTER 6

Swarms

A swarm of insects may be defined as a cluster that moves about and rests as a whole, keeping together by means of mutual attraction and stimulation. Individuals behave gregariously — locusts and bees are examples that immediately come to mind. The word 'swarm' is also used for clouds of insects that become clustered more because the individuals have some common activity than because they are mutually attracted or stimulated. Such clusters are more like milling throngs than true swarms (Oldroyd 1964). We look first at swarms of locusts.

6.1 SWARMS OF LOCUSTS

Locusts are certain species of short-horned grasshoppers that behave gregariously from time to time; otherwise they are solitary and behave like the majority of

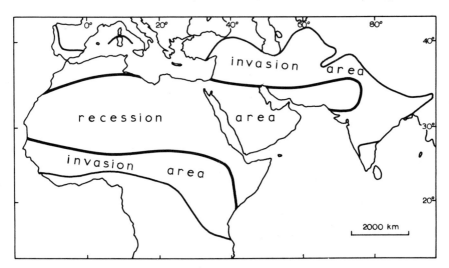

Fig. 104 — Invasion area of swarms of the desert locust, *Schistocerca gregaria*, during plagues and recessions.

grasshoppers. Whereas solitary locusts generally fly by night (page 162), swarms generally fly by day and they roost at night. Of all the dozen or so species of locusts, the **desert locust**, *Schistocerca gregaria*, has been by far the most studied. It is a worrying pest in Africa and south-west Asia, attacking cereals, sugar cane and other crops, as well as grazing. In times of plague it has been found at places within an area of some 30,000,000 km^2 in northern and eastern Africa and in south-western Asia (Fig. 104). Between plagues it is hemmed into a smaller 'recession area', where there may be few or no swarms. Plagues do not come at fixed times and although the causes are not fully understood, plague onset might be linked with the timing, amount and spread of the fitful rains within its often-dry 'invasion area', whereas a plague decline seems to be set biologically (Waloff & Green, 1975). Plague swarms (Fig. 105) often have an area of some tens of square kilometres, and contain some tens of millions of individuals in a square kilometre. In flight, densities can reach 10 in a cubic metre but are often several orders of magnitude less (Fig. 106). Individuals form streams, each with remarkable uniformity of heading (Fig. 107). In a swarm as a whole there are all possible headings, but those flying above a few hundred metres often head downwind (Waloff 1972), and the topmost individuals can reach the upper part of the daytime convective layer of the atmosphere. Streams turn back into the swarm at its edge, so the swarm has great cohesion, even in the presence of turbulence over many days (Fig. 108). Because higher-flying locusts head downwind but turn back into wind on reaching the swarm's leading edge, and then descend and land, the swarm as a whole moves downwind. Rainey (1963) has demonstrated such movement very clearly by tracking individual swarms across country in East Africa. Fig. 109 illustrates some tracks and the winds blowing

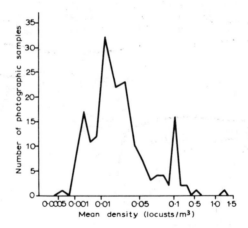

Fig. 106 – Frequencies of photographically-recorded mean aerial densities in swarms of immature desert locusts. (After Waloff 1972).

a   FAO photograph.

b

Copyright: Centre for Overseas Pest Research.

Fig. 105 – Swarms of the desert locust, *Schistocerca gregaria*.
  a Morocco, 1954.
  b Seen from an aircraft over Wajir district, Kenya, 13 January 1953 (area
    about 1 km²).

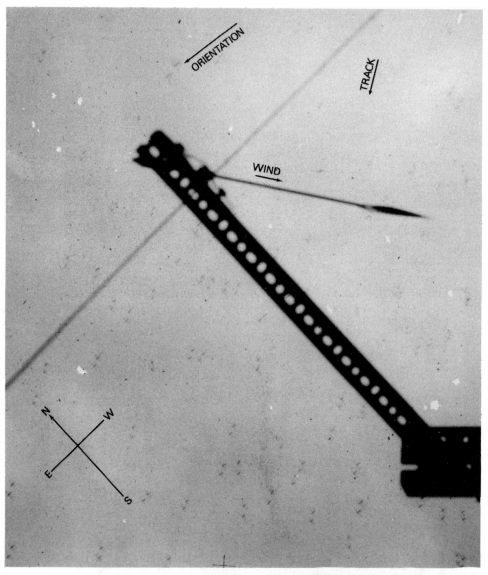

Fig. 107 – Looking vertically upwards into a swarm of desert locusts on the lower slopes of Mr Kilimanjaro, Tanzania, 1310 LT 22 February 1955. Double exposure with 1/50 second interval reveals tracks of individual locusts. Both tracks and headings are remarkably uniform but different, due to drift in the wind, whose direction is shown by the wind vane. Compare with Fig. 49.

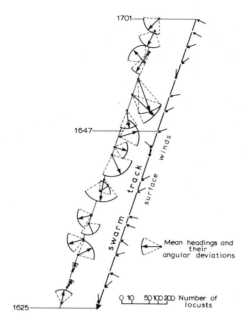

Fig. 108 – Variation with time of the headings of individual desert locusts flying at heights 3.6–18 m above the ground during the passage of the central part of a swarm near Kanga, Kenya, 10 February 1955; Headings measured photographically. Headings without a statistically significant mean shown without arrow heads. Mean surface wind 2 m/s. (After Waloff 1972).

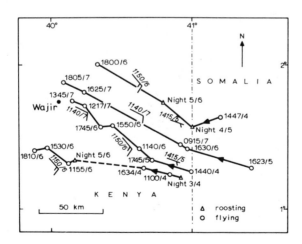

Fig. 109 – Examples of movements of warms of the desert locust near Wajir, Kenya, February 1954, tracked by aircraft and from the ground, showing downwind displacement. Winds at flying height; measured by balloon soundings at Wajir. One full feather represents 10 km/h. (After Rainey 1963).

at the same time, and Fig. 110 summarises the results of comparing swarm velocities with wind velocities for some 50 swarms. Tracks become complex when the windfield varies with time. Fig. 111 shows an example of a zig-zag track due to a marked diurnal variation of wind (page 45) near the north-facing escarpment in northern Somalia; and Fig. 112 shows an example of a looped track due to a reversal of wind over Sudan following a northward passage of the ITCZ (page 73). Some swarms have been tracked across country for up to a month, and there is circumstantial evidence that swarms can persist for several months. Daily flight duration in warm, sunny weather is typically about 10 h, leading commonly to daily displacements of 50-100 km. Biogeographical studies of long-distance migration by swarms of sexually immature locusts show that displacements of 1000-4000 km are not uncommon; they continue while

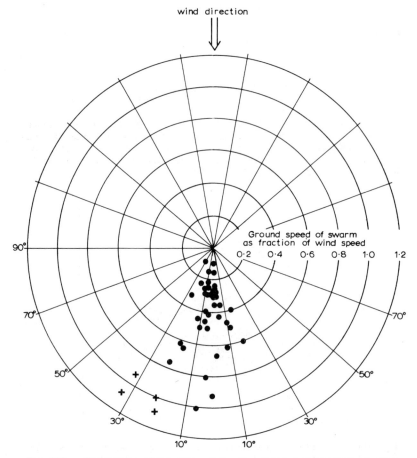

Fig. 110 – Comparison of velocities of swarms of the desert locust and winds measured at the time. East Africa, 1951-7. Crosses show swarms flying up to more than 900 m above the ground. (After Rainey 1963).

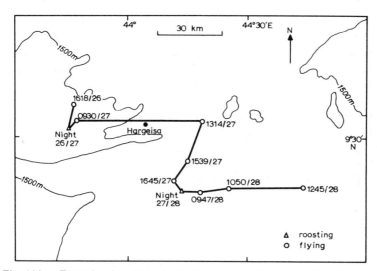

Fig. 111 – Example of a zig-zag track of a swarm of desert locusts in Somalia, September 1953. Due to diurnal variation of wind near a north-facing escarpment. (After Rainey 1963).

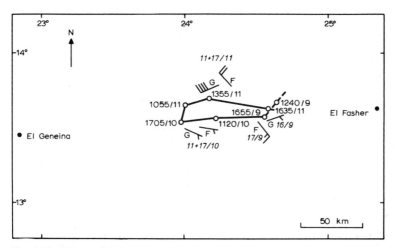

Fig. 112 – Example of a looped track of a swarm of desert locusts in Sudan, June 1955, due to northward movement of the ITCZ on the night of 10–11 June. (After Rainey 1963).

the locusts mature and lay their eggs. Downwind movement leads to migration of effectively the whole population between seasonal areas – migration that has been very well documented from countless swarm sightings over more than 40 years. Eggs are laid in rain-wetted soil from which they need to take up water before they can develop fully. For each seasonal rainfall there is often only one generation. Migrations from about May to July (from spring breeding at the

southern edge of middle latitude rains) are down the trade winds − blowing
from north or north-east over North Africa, and north or north-west over the
Middle East, Pakistan and India (Fig. 113). Over East Africa, migrations at this

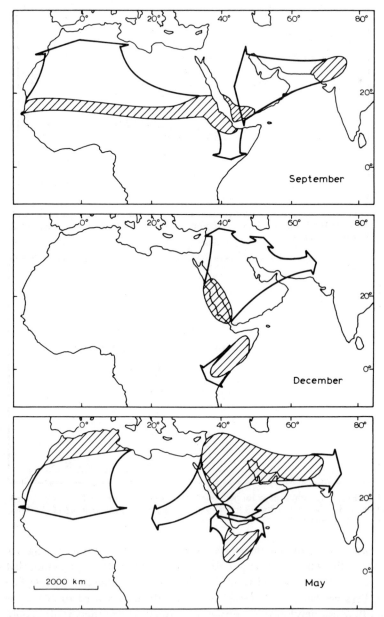

Fig. 113 − Swarming desert locusts − principal breeding areas (shaded) and
main migration directions of next generation adults (arrows). (After Betts 1976).

time of year are on southerly winds south of the ITCZ. Migrations from about
September to November are mostly away from the monsoon breeding areas —
westward from India and Pakistan to the Middle East, and from the sahel to
north-west Africa. Those are with easterly trade winds, but on some days there
is a movement northward with warm, southerly winds in the leading parts of
synoptic-scale wave disturbances moving eastward across North Africa and the
Middle East and sometimes taking locusts into Europe (Fig. 114). Over East
Africa, migrations at this time of year are on north-east trade winds north of the
ITCZ. Migrations from about December to April are away from the winter
breeding areas — northward from around the Red Sea, and south-westward

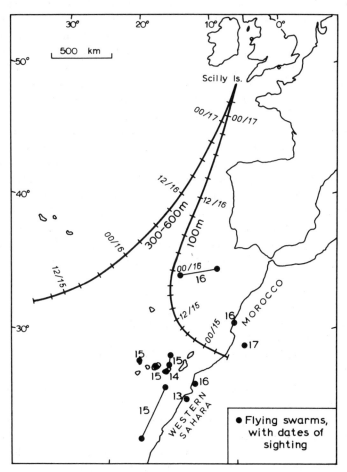

Fig. 114 — Back-tracks at heights of 100 and 300–600 m for desert locusts
arriving in the Scilly Islands at 0600 GMT 17 October 1954. The lower-level
back-track fits well the sightings of flying swarms at sea, in the Canary Islands
and along the African coast. (After Rainey 1963).

across East Africa. Numerous case studies of movements of whole populations of swarms in relation to winds over periods of days or weeks provide extensive and powerful circumstantial evidence supporting downwind movement (Rainey 1963, Pedgley 1981). Baker (1978), however, claims that these migrations result from flight mostly in preferred compass directions (much like the monarch butterfly, and other butterflies and moths – see page 95), and towards rain-in-sight and high ground, but Draper (1980) has confirmed that downwind movement is adequate to explain desert locust swarm migration. See also Rainey (1963) and Waloff (1966) for further extensive discussion of this species.

The **migratory locust**, *Locusta migratoria,* is another serious pest in the Old World. In Africa, for example, scattered individuals of the sub-species *L. m. migratorioides* are widely present south of the Sahara. Plagues start from out-break areas – habitats where *dry season* breeding is unusually successful, especially the middle Niger flood plain of Mali. Two good rainy seasons are needed for an outbreak to develop, during which scattered locusts move widely into areas outside the flood plain. In the first season there is a widespread build-up, but at low density. During the following dry season there is a concentration into the flood plain (see also page 122), and then early in the second rainy season, swarms begin to form. The last plague of swarms began there in 1928 and finished in 1941, reaching, at one time or another, most countries south of the Sahara (Fig. 115). Mapping the first few years of spread showed that swarms moved seasonally, reaching ever wider areas. They moved north-eastward from about March to September at the time of the monsoon south-west winds, and south-westward from about October to February at the time of the dry north-east trade winds. These to and fro movements suggest a link with seasonal changes in the position of the ITCZ, but few individual swarms were watched long enough to confirm downwind movement (Batten 1967 and 1972, Farrow 1974).

Another African species is the **red locust**, *Nomadacris septemfasciata.* Plagues of swarms again start from outbreak areas – certain grassy plains in inland drainage areas of eastern Africa, especially Tanzania and Zambia. The last plague lasted from 1930 to 1944, when nearly all of Africa south of the equator was infested at some time (Fig. 116). It almost certainly started from the plains of Mweru wa Ntipa, Zambia, and then an annual open downwind migration circuit was established (Symmons 1978) – north-west to Zaire and Angola, then south through Zambia, Zimbabwe and Botswana into South Africa. Unlike *Schistocerca gregaria,* swarms probably moved at most only a few kilometres a day, and because they flew low they were difficult to see against the woodland background. This species has one generation a year. Maturation occurs at the onset of seasonal rains; eggs develop in about one month, and hoppers in another two; and then adults stay immature throughout the rest of the rains and the following dry season. When numbers are great enough, swarms form and escape, but it seems that small swarms disperse and it is perhaps only

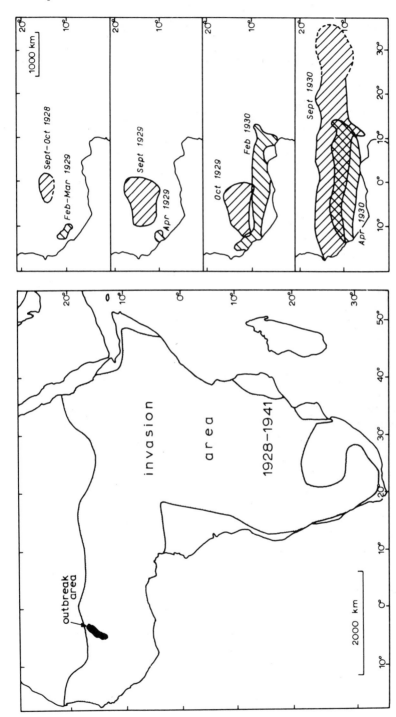

Fig. 115 — Outbreak and invasion areas of the African migratory locust, *Locusta migratoria migratoriodies*, for the 1928–41 plague. Seasonal expansion of the invasion area early in the plague are shown on the right. (After Batten 1967).

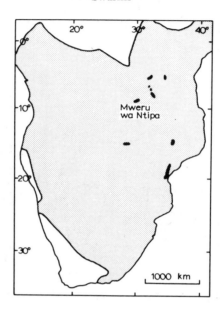

Fig. 116 – Outbreak and invasion area of the red locust, *Nomadacris septem-fasciata*, for the 1930–44 plague. (After Symmons 1978).

the largest that remain coherent enough to lay en masse and so set a plague under way.

The **Australian plague locust,** *Chortoicetes terminifera,* occurs over most of the continent, and outbreaks are common. It normally behaves as solitaries and flies by night (pages 126 and 146), but in plague years it can form day-flying swarms. These swarms are generally much less mobile than those of the desert locust; displacements are usually 10–20 km a day, and they last for only a few days (Casimir & Bament 1974, Farrow 1977), usually when spells of warm, dry, sunny north winds are blowing on the western sides of passing anti-cyclones. Another Australian species is the **spur-throated locust,** *Austacris guttulosa.* Like *Nomadacris septemfasciata* of Africa, it remains immature during the dry season. In September 1973, immature swarms invaded north-eastern New South Wales from Queensland, having been widespread north of 26°S since 1971. A new generation was produced on the summer rains, and swarms spread further south and east on north-westerly winds in October and November 1974, to reach the eastern coast in December (Casimir & Edge 1979).

Another African species is the **brown locust,** *Locustana pardalina.* It occurs mostly in the Karoo of South Africa, sheep-grazing steppe country and an outbreak region in which scattered locusts build up from time to time and from which swarms move out into an invasion area covering much of the southern continent (Lea 1964, 1969). This species resembles *C. terminifera* in its biology and ecology. Swarm movement is probably downwind, but this has not yet been demonstrated.

It may be noted that locust swarms illustrate the four main kinds of circumstantial evidence for long-distance and downwind movement by individuals of many other species: back-tracking to known sources, arrival with particular kinds of weather system, sightings far from sources, and seasonal reappearance after effectively complete clearance.

6.2 SWARMS OF OTHER INSECTS

In contrast to a swarm of locusts, which moves downwind, a swarm of **honeybees**, *Apis mellifera,* moves across country led by scouts that have already passed on information about possible new home sites. A swarm contains some 5,000 to 50,000 individuals with one queen; it occupies a volume of 15–30 m diameter and it moves at speeds up to 25 km/h. It will not go far without a queen; if it leaves without her, it will return (Morse 1963, Avitabile *et al.* 1975). A swarm of honeybees clearly moves within its boundary layer, and because of the swirling flight of individuals the swarm moves at some fraction of the speed of an individual. The distance moved is often a few kilometres at most, and seems unable to account for the apparent rapid rate of advance of the so-called Africanised bees in South America that resulted from the escape of some **African honeybees**, *A. m. adansonii,* from an apiary near Rio Claro, Brazil in 1957 (Taylor, O. R. 1977). This bee occurs widely south of the Sahara and seems to be most abundant in the central African plateaux where annual rainfall is 500–1500 mm, falling in a marked wet season. Its well-known but not understood migratory movements there would be an advantage in such a climate, and this ability may have been taken to South America. From the duration of waggle dances, Fletcher (1978) deduced that flight range is probably no more than a few kilometres, but Otis *et al.* (1981), from the amount of honey fuel carried by absconding swarms leaving their nest sites, estimated that flight could be up to 100 km. A succession of flights over days or weeks could account for the distances travelled. In Sri Lanka, the **giant honeybee**, *Apis dorsata,* also has a seasonal redistribution. Swarms from northern coastal plains fly to the hills in June–July and return in December — a distance of 150–200 km undertaken in short steps with resting and feeding spells lasting a few days (Koeniger & Koeniger 1980). Although these movements take advantage of seasonal flowering periods, they are against the seasonal winds, and are apparently genetically fixed.

'Swarms' of midges are well known. Blood-sucking **midges**, *Culicoides,* often gather in dancing swarms of males into which females fly to find a mate. Field observation of *C. nubeculosus* by Downes (1955) showed that each individual always heads upwind and by sight keeps a more or less fixed position in relation to a marker — a dark or light object that stands out against its surroundings. Because the midges' air speed is only about 1 m/s, gusts carry them temporarily downwind, and in all but the gentlest breezes they seek the lee side of a shelter, or they land. Likewise, *Culicoides brevitarsis,* a possible vector of the virus causing **bovine ephemeral fever** (see page 139), can gather over a marker in winds of 1 m/s but not 2 m/s (Campbell & Kettle 1979). Larger markers have larger and higher-flying swarms. High-speed cine photography of swarms of the **gall midge,** *Anarete pritchardii,* has shown that each individual flies more or less independently, in contrast to locusts in a swarm, although there is avoidance of collision (Okubo *et al.* 1977). Movement is in a sequence of roughly straight flights at 30–40 cm/s lasting about 0.1 s, with sharp turns up to 180°, equally to left or right (Goldsmith *et al.* 1980), and with accelerations of 2 *g* not uncommon, and reaching 5 *g*. Okubo *et al.* (1981) have developed a photographic technique for measuring the three-dimensional structure of such 'swarms' by using the positions of both the insects and their shadows. Other biting flies, such as mosquitoes, horse flies (tabanids) and black flies (simuliids), also form male swarms for mating, and these species are well known for their persistence as threatening clouds around a prey (Downes 1969). Loops, zig-zags and figure eights are common tracks of individual males — presumably to monitor the 'swarm' edges for arriving females (Sullivan 1981). A cloud of **flies** of unknown species seen by Paterson (1975) from the rim of the gorge of the River Tarn, in France, demonstrated the presence of a lee eddy (Fig. 117). From time to time the separation point moved a little upwind, temporarily immersing the rim in the cloud. As the separation point retreated, most flies returned to the gorge, but some remained in the lee of a tree, where light winds presumably persisted.

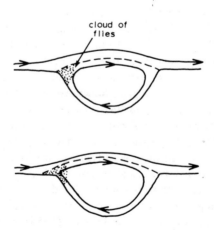

Fig. 117 – Relation between the position of a cloud of flies and movements of the stagnation point of a presumed standing eddy in the gorge of the River Tarn, France. (After Paterson 1975).

There have been many reports of 'swarms' of day-flying butterflies; often they recur seasonally and are parts of large-scale migrations. Night-flying moths also occur occasionally in 'swarms'. Brown *et al.* (1969) recorded such an occurrence of the **African armyworm moth,** *Spodoptera exempta,* along 55 km of road north-west of Nakuru, Kenya. It was part of a major redistribution of population moving north-eastward along a front of 600 km, judged by a sudden change in the area of egg laying over a 3–4 day period, as well as by light-trap catches.

CHAPTER 7

Dispersion and concentration

Some organisms take off together in vast numbers. The resulting cloud may be dense and easily seen, as with some insects and seeds, but often it is thin and hard to find, especially with small organisms such as viruses, fungal spores and pollen grains. As it moves across country, a cloud tends to be dispersed, not only because some individuals land (Chapter 2) but also because the remainder tend to be spread further apart, and so the cloud becomes thinner. Despite the ever-present tendency for a cloud to disperse, surprisingly little work has been done to understand the mechanisms and effects of **dispersion.** We first consider the dispersion of minute windborne particles — smoke; then we look at organisms and the effects of fall through the air; and lastly we consider insects, whose flight can at times lead to **concentration,** the opposite of dispersion. (For a thorough discussion of dispersion see Pasquill 1974.)

7.1 CLOUD STRUCTURE

Some idea of the complexity of the dispersion process can be got by watching smoke, although it should be remembered that the buoyancy of warm, smoky air from a fire aids dispersion. The **source** may be a *point,* from which smoke can drift either as a puff or a plume. A puff is produced more or less instantaneously, as with an explosion; a plume, as from a chimney, may be looked upon as a continuous series of overlapping puffs. Alternatively, the source may be a *line* or an *area* along, or within, which there are many point sources with various intensities; a forest fire is an example of the latter. In all cases, smoke disperses as it streams downwind. The cloud is patchy, and its patchiness is always changing. Dispersion may be thought of as taking place in the following way. Because the air is always being more or less churned up by eddies, leading to its characteristic gustiness, threads or sheets of clear air are drawn into the cloud at the same time as threads or sheets of smoky air are pulled out into the clear air. In this way the cloud is teased apart, and it therefore becomes less dense as it grows in size. Eddies on the edge of a cloud, because they bring in clear air, tend to increase

the patchiness. By contrast, eddies within the cloud tend to mix it and decrease
the patchiness. A dispersing cloud therefore tends to be least dense, on average,
at its edge, and the chance of finding a clear patch decreases inwards.

On a sunny day, when convective mixing extends to hundreds or thousands
of metres above the ground, smoke is quickly diluted and carried aloft (Fig.
118a). Much the same is true for sunless but windy days, when ground roughness
alone produces the eddies. In both cases, the cloud top will be more or less
clearly defined by the top of the mixed layer, where there is often a change of
lapse rate. By contrast, within a temperature inversion, where eddy mixing is
greatly reduced and may even almost disappear, a smoke cloud drifts with little
mixing, as can be seen on clear, quiet evenings (Fig. 118b). For intermediate
cases, where there is no inversion but the temperature lapse rate does not allow
convection, mixing does take place but it is less vigorous than on a sunny or
windy day. Because there are strong diurnal variations of lapse rate near the
ground, particularly in sunny weather, the intensity of mixing decreases rapidly
after mid afternoon with dramatic effects on the intensity and depth of the
mixing — smoke from an evening fire tends to cling near the ground in clear,

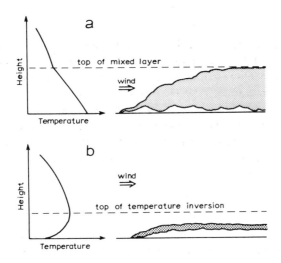

Fig. 118 — Schematic variation of plume shapes from a point source in the
presence of: (a) a well-mixed layer, (b) a temperature inversion.

quiet weather. Mixing is stronger over rough country — hills and cities — rather
than over flat country. Much has yet to be learnt about the nature and origins
of plume structure over distances of hundreds of kilometres, although there have
been some measurements from aircraft of the concentrations of tracers such as
particles, chemicals and radioactive materials.

A consequence of mixing within a smoke cloud is that any two particles released close together tend to drift apart. This can be demonstrated more clearly with soap bubbles blowing in the wind, or with pairs of fixed-level balloons tracked by radar. Order of magnitude separations are 100 m after 1 min, 300 m after 10 min, and 3 km after 100 min (the last being the equivalent to a 50 km downwind drift in a 10 m/s wind).

When there is vertical wind shear, a smoke cloud becomes distorted and may lead to the formation of layers aloft. Strong shears are present in coastal and mountain wind systems (page 52), and it is in them that cloud structures can become very complex.

7.2 CLOUDS OF POLLEN AND BACTERIA

It seems likely that clouds of small organisms behave rather like smoke, once allowance is made for settling under the influence of gravity. Little work has been done, however, on the dispersion of such clouds, except for the time-integrated effects on vertical profiles of density and on deposition.

Density profiles are often given in the form

$$\ln Q = \ln Q_0 - \frac{w}{a} z \tag{1}$$

where Q is the density at height z above the ground, Q_0 is the density at the ground, w is the fall speed of the organisms through still air, and a is a coefficient that expresses how well the air is being mixed, and that varies with occasion. Such an equation is to be expected where mixing has resulted in a steady state in which the net upward flow rate due to eddies is just balanced by fall under gravity, and where the rate of mixing does not vary with height — as is likely to be so above the well-mixed layer or at night. If the rate of mixing increases with height, as is likely to be so near the ground by day, then the equation is likely to be of the form

$$\ln Q = \ln Q_0 - \frac{w}{b} \ln z \tag{2}$$

where b is averaged over the profile depth and is a constant on a particular occasion, of order 10^{-2} m/s. In practice, a steady state is unlikely to be reached because, for example, there has been insufficient time, or the take-off rate has been varying, or there have been many discrete sources rather than one large and continuous source.

Raynor et al. (1966) introduced a technique of using artificially coloured pollen grains to examine the behaviour of **pollen clouds**. By using circular plots of timothy grass, *Phleum pratense*, surrounded by fallow having grids of samplers on masts up to 4.6 m above ground, they showed that mean concentration had fallen to less than 10% of that at 1 m by a few tens of metres from the edge of the source (Raynor, Ogden & Hayes 1971; see Fig. 119). Similarly with line

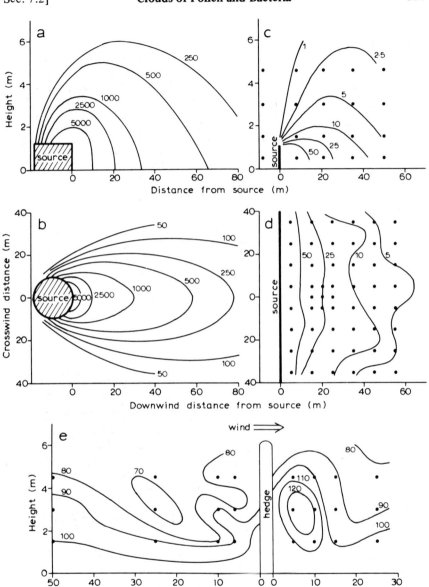

Fig. 119 – Typical patterns of time-averaged pollen concentrations (grains/m³) in plumes downwind of various sources: horizontal and vertical (centre-line) planes. Dots show positions of samplers.

(a) and (b): timothy grass pollen from an area source; the horizontal pattern is for a height of 1.5 m. (After Raynor, Ogden & Hayes 1971).

(c) and (d): castor bean pollen from a line source; the horizontal pattern is for a height of 0.5 m. (After Raynor, Ogden & Hayes 1973).

(e): mixed tree pollens from distant sources; concentrations normalised to a reference position 50 m upwind at a height of 1.5 m. (After Raynor, Ogden & Hayes 1974).

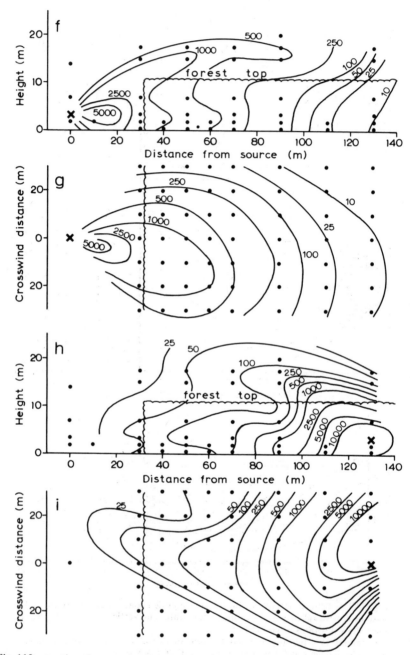

Fig. 119 –*continued*
(f) and (g): ragweed pollen from a point source (x) at a height of 3.5 m and
30 m upwind of a forest edge. (After Raynor, Hayes & Ogden 1974b).
(h) and (i): ragweed pollen from a point source (x) at a height of 3.5 m and
100 m upwind of a forest edge. (After Raynor, Hayes & Ogden 1975).

sources of 20-30 μm diameter grains from castor bean, *Ricinus communis*, giant ragweed, *Ambrosia trifida*, and summer cypress, *Kochia suparia*, it was found that concentrations agreed with calculations based on a Gaussian diffusion model (Raynor, Ogden & Hayes 1973). Further experiments near dense hedges showed that the most concentrated part of the pollen cloud, near the ground, is raised over the hedges, and lee-side eddies then lead to densities greater than would have occurred in the absence of the hedges (Raynor, Ogden & Hayes 1974a). The effect of a coniferous forest 12 m tall were examined by using stained *Ambrosia* pollen dispersed in minute water droplets that soon evaporated (Raynor, Hayes & Ogden 1974b). The plume was found to broaden both up-wards and sideways as it approached the forest edge, the sideways spreading continued within the first 60-80 m of forest, beyond which vertical mixing became uniform and cloud density fell off downwind much faster than in open country. Part of the cloud flowed above canopy top. Deep within a forest, cloud drift can be very complex, judged by both smoke (Oliver 1973, 1975) and pollen plumes, which become curved and split because of the patchiness of foliage density (Raynor, Hayes & Ogden, 1975). On a larger scale, aircraft flights in sea breezes revealed pollen being carried upward at the leading edge of the sea air (Raynor, Hayes & Ogden 1974c).

From studies of short-range dispersion downwind from isolated trees (Wright 1953), or medium-range from woodland (Tampieri *et al.* 1977), it has been shown that

$$\ln Q = \ln Q_0 - \frac{x}{d} \tag{3}$$

where d is a dispersion distance, of order 100 m. It follows that this is the minimum spacing needed to be left between scattered trees after regenerative cutting of forest. Moreover, seed orchards need to be further apart than this from isolated trees to adequately reduce contamination (Wang *et al.* 1960).

7.3 CLOUDS OF INSECTS

Several trapping studies of downwind drift by **scale insects** (page 89) have shown a rapid fall of time-averaged crawler catch with increasing distance from source, x, usually in a relationship like

$$\ln Q = \ln Q_0 - \frac{x^{1/2}}{e} \tag{4}$$

where e is a constant on a particular occasion. Willard (1974) set up a line of sticky traps from the edge of an abandoned lemon orchard near Adelaide, Australia, and found that **Californian red scale**, *Aonidiella aurantii*, was caught out to at least 300 m (Fig. 120). Stephens & Aylor (1978) used sticky traps to catch **red pine scale**, *Matsucoccus resinosae*, in a line downwind of a stand of red pine,

Pinus resinosa. Suction traps have been used to catch **beech scale,** *Cryptococcus fagisuga,* downwind from infected trees (Wainhouse 1979). Although numbers caught decrease rapidly downwind, the outward flux of insects (the rate at which they cross a circle centred on a point source) may vary little with distance, showing that few individuals are lost near the source, and many more reach greater distances than might otherwise be expected. For example, McClure (1977) used four lines of sticky traps from an isolated small stand of eastern hemlock, *Tsuga canadensis,* in Connecticut, to catch crawlers of **elongate hemlock scale,** *Fiorinia externa.* Catches showed that flux was about constant to at least 105 m.

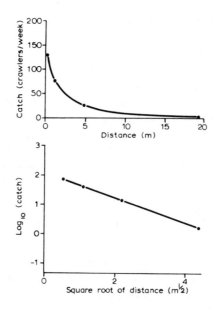

Fig. 120 – Average weekly catch of California red scale crawlers, *Aonidiella aurantii,* December 1966 to June 1967, and its variation with distance from source. Untransformed data above; transformed data below, showing a linear relation between $\log_{10}$ (catch) and the square root of distance. (After Willard 1974).

In contrast to wingless insects, which have little control over their landing, the dispersion of insects in flapping flight tends sooner or later to be affected by purposive landing (Taylor & Taylor 1977, Taylor 1978); hence the variation of cloud density with distance from source is more complex than for organisms that simply drift. Most generally, it can be expressed as

$$\ln Q = A + B \ln x + C x^D \tag{5}$$

where the four parameters, A to D, are constant on a particular occasion, but they vary with species, time and place (Taylor *et al.* 1979). One or more may be zero; for example, (5) reduces to (4) if $A = C$, $B = 0$, $C = 1/d$ and $D = \frac{1}{2}$.

Where insects carry plant viruses, the variation with distance from virus source of the incidence of subsequent plant disease is often represented by a similar expression. The infection gradient, however, is likely to be steeper than the deposition gradient because not all transmissions of virus lead to disease (Gregory 1973, Thresh 1976). With strongly flying vectors and persistent virus, infection gradients are shallow, and diseased plants are scattered over a wide area. By contrast, with weakly flying vectors and non-persistent viruses, infection gradients are steep, and diseased plants are clumped near initial foci. In either case, however, peripheral infections may be more numerous than the denser and more obvious infections near the source, and they are consequently of greater importance as potential secondary sources. Gradients are steeper from point than from area sources, and the larger the source the greater the chance that individual insects can take virus over greater distances, due to variability of behaviour within both the insect population and its environment. Initial foci within a crop may be due to randomly distributed crop or weed plants, or to random settling of infective vectors coming from afar. But subsequent spread of disease may be due to a vector that is a less active species, or even a less active form of the same species that first brought the disease.

Vertical profiles of airborne insect density, averaged over spells of an hour or more, can often be written in the form

$$\ln Q = \ln Q_0 - f \ln\left(\frac{z}{g} + 1\right) \tag{6}$$

where f and g are constant on a particular occasion (Johnson, C. G. 1957). Because (6) is the same as (2) when z is very much greater than g (and $f = w/b$) it may well be that profiles like those given by (6) represent steady-state clouds in which an upward flux of insects due to eddy mixing balances a downward flux due to fall out or to active flight. Near the ground, where z is about as small as g, cloud densities given by (6) become less than those given by an equation like (2). This is not too surprising when it is recalled that flight near the ground is likely to be strongly affected by goal seeking that leads to landing. It seems likely that profiles such as those given by (6) are more typical of flight above the insect boundary layer than below. It should also be remembered that these profiles are *time-averaged;* instantaneous profiles can be markedly different, particularly in the presence of vertical wind shear. Moreover, insects may select one or more heights at which to fly on a given occasion determined, for example, by temperature or apparent ground speed (page 165). In this way, layers can develop within which cloud density is greater than above or below. Such layers may not be visible from the ground but have been well seen by radar — for example, moths in Canada (Schaefer 1976), mixed insects in East Africa (Riley *et al.* 1981; see Fig. 121), grasshoppers in West Africa (Riley & Reynolds 1979), mixed insects over the River Niger, West Africa (Reynolds & Riley 1979; see Fig. 129), and locusts in Australia (Reid *et al.* 1979). Quantitative measurements of volume density can be made by radar, both from the ground (Fig. 122) and from an aircraft (Fig. 123).

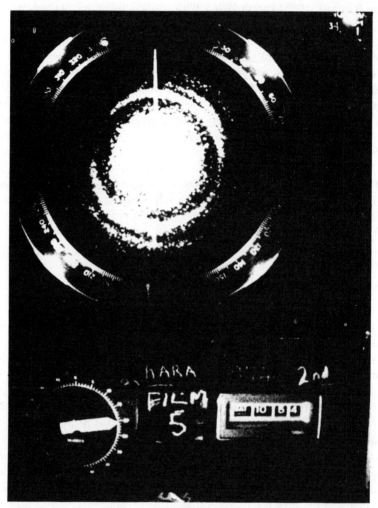

Copyright: Centre for Overseas Pest Research.

Fig. 121 – Flying insects seen by radar at Kara, Mali, 2254 LT 2 November 1974. Range 6 km, elevation 20°. Apart from insects near the ground (the central echo mass), there are well-marked layers (ring echoes) at heights of about 1200 m, 1400 m and 1700 m. The ring echoes show considerable uniformity of heading (cf Fig. 102). (After Riley & Reynolds 1979).

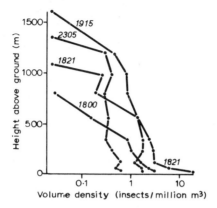

Fig. 122 – Vertical profiles of volume density of flying grasshoppers. Measured by ground radar at Radma, Sudan, 21 October 1971 and showing variation with time. Mainly *Aiolopus simulatrix*. (After Schaefer 1976).

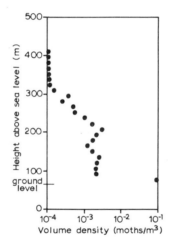

Fig. 123 – Vertical profiles of volume density of flying spruce budworm moths, *Choristoneura fumiferana*. Measured by downward-pointing radar from an aircraft over New Brunswick on the evening of 10 July 1976. (After Greenbank *et al.* 1980).

7.4 CONCENTRATION OF WINDBORNE INSECTS

In a cloud of spores, pollen, wingless insects or other organisms simply drifting with the wind, volume density is controlled largely by source strength and behaviour of the atmosphere. By contrast, the volume density of a cloud of winged insects *in flapping flight* can be strongly affected also by insect behaviour.

For example, massive and simultaneous take-off or landing can clearly lead to large changes in volume density; so, too, can a tendency to stay above a particular place once many individuals have reached it. This can lead to an accumulation near a marker, as with midges (page 181), or in the lee of an object that gives shelter from the wind. Such accumulations can sometimes be seen by eye, and volume density profiles, in both vertical and horizontal, can be measured more or less accurately with suction traps.

Most **crop shelter** on farms in temperate Europe and North America is given by hedges or rows of trees, but artificial windbreaks made of, for example, cane, straw or netting are widely used to shelter small areas of valuable crops. Lewis (1965b) demonstrated that *flying* insects tended to gather near a windbreak, especially on the downwind side, and that horizontal profiles of volume density are similar to those for *drifting* organisms like spiders, or even for pieces of paper (Fig. 124). Although this suggests little behavioural effect by many species on their concentration whilst windborne, it is likely that some do respond to the weaker wind and greater gustiness on the leeward side of a windbreak. For example, night-flying **lacewings** and **moths**, tend to gather closer to windbreaks than do day-flying **thrips** and parasitic **wasps**; and so do strong fliers compared with weak ones (Lewis 1968). Volume density is affected as far downwind as ten times the height of a windbreak, but greatest densities (up to ten times those upwind) are mostly at distances one to four times the height of the windbreak (Lewis 1969b). In further field studies, a 300 m north-south belt of trees, mostly pine about 20 m tall and 12–14 m wide at the top (but only

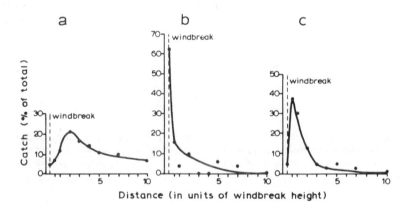

Fig. 124 – Distribution of airborne insects downwind of a windbreak. (After Lewis 1970).
    (a) aphids blown from upwind
    (b) leafhoppers (jassids) blown from the windbreak
    (c) dung flies (sphaerocerids) blown from beneath the windbreak and from upwind.

8 m at the base, where there was much hawthorn between the tree stems), was used on 16 days with westerly winds during June and July 1969 to discover how airborne insects gathered behind shelter (Lewis 1970). There was a sheep pasture on the windward side, and a wheat field to leeward. Suction traps were put in the field along a line out from the trees. The kinds of insects in the trees were found by beating and, in the wheat ears, by taking samples along lines through the traps at fixed distances from the trees. The belt was 34% open, leading to greatest shelter (60% — found by a line of sensitive cup anemometers) at 40 m from the belt. Insects blown on the wind from elsewhere (such as **aphids**) had a profile peaked in much the same place as the amount of shelter, perhaps because they had been blown over the trees and down to leeward. Insects blown from the trees (such as **leafhoppers**) had their greatest density close to or within 40 m of the trees. Faster-moving **dung flies** from beneath the trees and to windward had a sharp peak in the density profile at 10-20 m. The downwind distribution of **cereal thrips** in early July almost certainly came about from the pattern of flight in early June, when the winged females would have left their wintering sites in grass, litter and bark to be blown by the wind on to the wheat, where they would have laid their eggs and become less willing to fly. First females are sparse and not easily visible, unlike their numerous male offspring, which are wingless, and stay on the plants where they hatch.

Living windbreaks such as hedges and tree-belts give not only shelter to crops but also feeding and breeding places for insects, some of which can be predators or parasites of those that are brought to the crop on the wind. The downwind density variation of insects coming from afar is then changed by insects coming from the hedge or tree-belt. An artificial barrier, however, may well encourage field pests to gather without being able to harbour their hedge-living enemies, but it is not yet known if windbreaks are in general harmful or useful. Each windbreak and crop must be judged individually until more is known of the ways by which they gather insects.

Temporary aggregation of **ladybird beetles**, *Hippodamia convergens,* and *H. tredecimpunctata,* on lake shores in the USA has been attributed to the beetles falling into the water during migration and then being washed ashore (Lee 1980), but it could also have been due to settling on the first available land. Concentrated landing by pest species accumulating in sheltered areas may greatly increase the harm they cause (page 62).

The occurrence of rain may also increase volume density by inducing insects to fly nearer the ground; this would help explain increases in trap catches during rain.

Snow (1979) used suction traps at seven heights up to about 8 m above the ground at a site near Bansong, the Gambia, to measure vertical density profiles of several species of mosquitoes, and found that flight became concentrated near the ground as wind speed increased, perhaps in an attempt to keep angular velocity constant (page 165).

Mass take-off from a restricted area gives a cloud in the form of a plume (Fig. 125). Such plumes have been seen by radar with **African armyworm moths**, *Spodoptera exempta*, flying away from the site of an earlier caterpillar outbreak (Riley *et al.* 1981) and with **cotton bollworm moths**, *Heliothis armigera*, from a field of groundnuts in the Gezira of Sudan, where they had been feeding on nectar (Schaefer 1976).

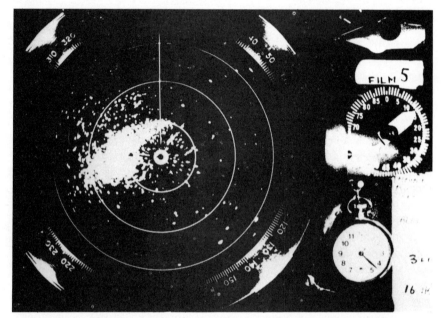

Fig. 125 – A plume of moths shown by radar streaming downwind towards 250°) from·the site of an outbreak of African armyworms. *Spodoptera exempta*, 45 km north-west of Nairobi, Kenya, 0122 LT 3 April 1979. Range rings 460 m apart; elevation 14°. (After Riley *et al.* 1981).

Light traps are widely used at night to indicate qualitative changes in volume density, but records are more difficult to interpret than those from suction traps because they are affected by other light sources, for example, varying timing and intensity of moonlight (Bowden 1973b, Bowden & Church 1973, Bowden & Morris 1975), and by changes in wind speed (for example, Douthwaite 1978). Comparing catches from traps at different heights in the same place reduces these difficulties. Observations by Taylor *et al* (1979) at Muguga, near Nairobi, Kenya, using traps at ground level and 24.5 m to sample a mixed population of night-flying moths showed that larger species tend to fly higher, irrespective of family. This is consistent with observed downwind flight there (page 162) because small and therefore generally slow-flying moths do not need to fly as high as larger species to have the wind dominate their flight direction. Pheromone traps also have the disadvantage of catch being affected by wind speeds greater than the air speed of the insects being caught.

Increased volume density can also be brought about by **wind convergence** — the net inflow of air into a particular volume. For practical purposes, three-dimensional convergence is zero; hence horizontal convergence must be compensated by vertical divergence. One consequence of horizontal wind convergence is a shrinking with time of the area within which a particular cloud of insects is flying. Another consequence is a tendency for insects to be carried away in the vertical divergence. Near the ground, vertical divergence must necessarily be upward, leading to an upward component of the wind that increases with height. Hence horizontal wind convergence in insect-laden air near the ground tends to be accompanied by an upward flux of insects. But if this flux is limited by insect behaviour, for example, by reaction to changes with height of temperature or wind, then cloud depth can remain more or less constant as cloud area shrinks — that is, cloud volume decreases and volume density increases. Some idea of the rate at which windborne insects can be concentrated in this way can be found from the definition of horizontal wind convergence, C, as the rate of shrinking of unit area with time:

$$C = -\frac{1}{A}\frac{dA}{dt} \tag{7}$$

Let t_c be the time needed for a tenfold increase in density. Integration of (7) gives

$$t_c = \frac{2.3}{C}$$

Windfield maps show that C is of order $10^{-5}/s$ for synoptic-scale wind systems (whence t_c is of order $2.3 \times 10^5 s$, that is, a few days); for mesoscale systems, such as coastal and rainstorm winds, it is of order $10^{-4}/s$ (whence t_c is about 10 h); and for small scale systems, such as simple convection cells, it is about $10^{-3}/s$ (whence t_c is about 1 h). Within these systems there will be parts, especially at windshift lines, where convergence is stronger by an order of magnitude or more. Thus, t_c is much the same as the life span of the wind system itself, no matter what its size may be. A single wind system can therefore be expected to bring about something like a tenfold increase in volume density, as long as the flying insects remain within the converging air. This rate is comparable with that due to insect behaviour within a region of shelter (see above). These theoretical estimates of concentration rate within converging air have not yet been confirmed by field observations — not surprisingly, because it has not yet proved possible to follow a given cloud of insects whilst it is being concentrated.

Sudden increases in insect numbers (volume density) have for long been associated with windshift lines. There are several reasons why this should be so:

change in source with change in wind direction; triggering of take-off with change in gustiness or light intensity or some other property of the windshift; concentration of flight nearer the ground with onset of rain; accumulation of individuals by horizontal wind convergence. Field studies using radar with **locusts** in Australia (Reid *et al.* 1979), with **moths** in both Canada (Greenbank *et al.* 1980) and East Africa (Riley *et al.* 1981), and with **grasshoppers** in West Africa and Sudan (Schaefer 1976) have shown that marked echo bands hundreds of metres broad accompany some windshift lines associated with sea breezes and rainstorm outflows (page 53, see Fig. 126). The evidence suggests that flying insects accumulate in the zone of convergence, particularly in the strongest

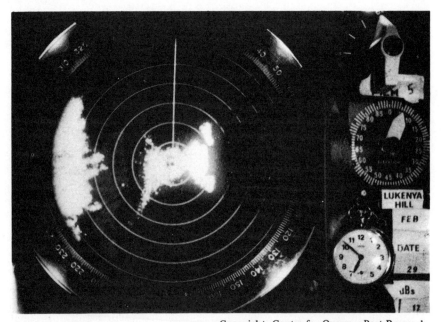

Copyright: Centre for Overseas Pest Research.

Fig. 126 – Flying insects forming a band echo along a windshift line at the leading edge of an outflow from rainstorms (the dense echoes to the west; those to the east are from a hill). 1852 LT 29 February 1980, at a site 35 km south-east of Nairobi, Kenya. Range rings 1850 m apart; elevation 6°. As the band passed overhead it could be seen to consist of countless individuals converging into the windshift from both sides.

part where there is an overturning of the lower atmosphere. In Fig. 127, insects within the inflow coming from the right are taken aloft near the leading edge of the cool air and then tend to be carried away in the outflow aloft. If, at the same time, insects actively fly downwards they re-enter the inflow and are recirculated. In this way, volume density increases near the leading edge, the rate depending on the fraction that recirculates. Symmons & Luard (1980) have attempted to quantify this process of concentration by means of a numerical model.

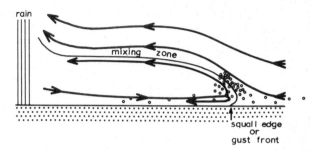

Fig. 127 – A possible mechanism for concentrating flying insects at a meso-scale windshift line due to recirculation following downward flight from the zone of ascending air.

When such concentrations of insects land they may lay eggs that lead to dense infestations of the next generation – perhaps one or two orders of magnitude greater than nearby. Such mass layings may be the cause of some of the outbreaks of the **African armyworm**, *Spodoptera exempta* (page 207) and also of the variability in distribution of eggs of the **cotton bollworm moth**, *Heliothis armigera,* in the Gezira of Sudan (Haggis 1981).

Active downward flight may account for the puzzling rarity of dead insects within hailstones, for the stones form within air scooped up by the advancing squall from the storm (Browning 1981). If the outflow is so far above the ground that temperatures there are too low for flight, insects will not be able to enter it (unless upcurrents are too strong to allow fall-out) and accumulation will be enhanced. Insects approaching from the left could also become concentrated in the same way after flying into the cool inflow. An approximately tenfold increase in *volume* density seems likely on both observational and theoretical grounds. Moreover, if there is subsequently a mass landing, a further approximately tenfold increase in *areal* density on the ground near the windshift as compared with the inflow seems possible because of the much greater depth of the overturning cloud. Such a concentration at sea breeze fronts may have accounted for some of the annoying clouds of **rose-grain aphids**, *Metopolophium dirhodum,* over south-east England in July 1979 (Cochrane 1980).

A similar concentration of *wingless* insects can be expected, as long as they fall fast enough. For example, caterpillars of the **gypsy moth**, *Lymantria dispar,* drifting on silk threads (page 92), have been described as being concentrated by sea breeze fronts (Cameron *et al* 1979). On a smaller scale, Schaefer (1976) and Reid *et al.* (1979) report radar observations of insects being concentrated in the walls of polygonal convective cells (Fig. 128). In East Africa **armyworm moths**, *Spodotera exempta,* have been seen concentrated in rotors (Riley *et al.* 1981).

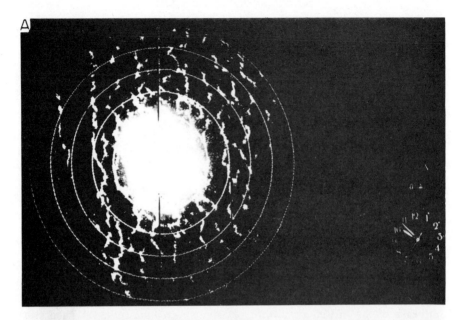

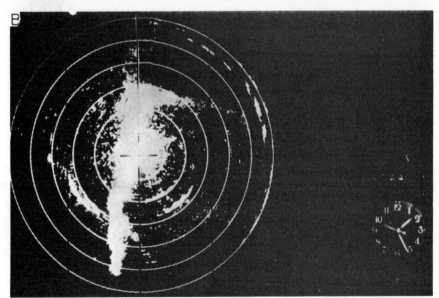

Fig. 128 – Insects concentrating at the boundaries of polygonal convection cells at Coonamble, New South Wales, 1100 LT 14 March 1971. (a) Range rings 1850 m apart. (b) Corner of a cell – dragonflies and butterflies.

Thus, windshifts associated with coasts and rainstorms seem able to bring about an increae in *areal* density by some two orders of magnitude; they act like brooms in the atmosphere that sweep windborne organisms into 'heaps' that may lead to serious outbreaks of pests and diseases. Synoptic-scale convergence, by contrast, seems unlikely to be as effective in producing such great concentration directly because updraughts are weaker and flying insects would generally be able to avoid being taken aloft, thereby preventing the deepening of the cloud.

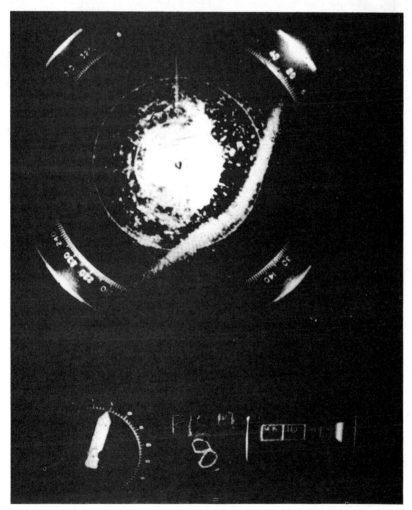

Fig. 129 − Flying insects, probably mostly Ephemeroptera, concentrating over the River Niger and shown by radar at Kara Mali, 2239 LT 4 November 1974. Range rings 460 m apart; elevation 4°. (After Reynolds & Riley 1979).

Indirectly, however, synoptic-scale convergence may be effective because of its tendency to be associated with rainstorms. There is no evidence that wind convergence adds significantly to insect behaviour in the production and maintenance of locust swarms (page 169).

Band echoes from night-flying insects over the River Niger in Mali seem to be caused by insect behaviour, using the river as a marker (Reynolds & Riley 1979, see Fig. 129), but on a larger scale, night-time land breezes converging on the same river during the flood season seem to concentrate the **African migratory locust,** *Locusta migratoria migratorioides,* into new breeding areas as the surrounding arid region dries out (Davey 1959).

So far, almost all the evidence supporting concentration of airborne insects by convergent winds has been circumstantial. Nevertheless the implications are considerable (Rainey 1976). For example, the process of concentration may bring individuals within range of mutual perception, especially when there is continual internal movement so that individuals pass close to each other. Crowding together may be significant for population dynamics and for crop damage, and it may be sufficient to make aerial control against airborne insects worthwhile.

CHAPTER 8

Forecasting

It is becoming ever more widely recognised that efficient pest management depends upon having a forecasting service that can advise on where, when and by how much the pest population is likely to grow and spread. Of particular value is an ability to forecast changes that are likely to cause harm beyond an economically acceptable threshold. Time, money and other resources can be needlessly wasted without such advice; or untimely, insufficient and even harmful measures can be applied in attempts to keep pest numbers below an acceptable level. Much the same is true of disease organisms. Forecasting pest and disease development is therefore an essential part of their management, just as much as forecasting crop yield. For discussion of the role of forecasting in pest and disease management see, for example, Pimental (1975), Scott & Bainbridge (1978), Horsfall & Cowling (1978), Jones & Clifford (1978), Ebbels & King (1979) and Thresh (1981).

8.1 NEEDS OF FORECASTING SERVICES

The preparation of any forecast depends upon knowing both the **present state** and the **causes of change**. The latter can be expressed in the form of a more or less complex model that includes the effects of the changing environment on the initial state. A great variety of empirical models, many very simple, has been developed for forecasting pests and diseases. In recent years, some much more complex numerical models have been devised, based on understanding of quantified relationships between organism and its host, as well as on the constantly varying environment. These models have the advantage of showing up the gaps and errors in our understanding of the population dynamics of the pests, and of the epidemiology of disease organisms. One of the most serious of these gaps is the role of **mobility**, including windborne mobility. Some models simulate reality adequately without taking any account of mobility, but only if the initial population is well described. Mobility must be taken into account, however, if outbreaks of pests and diseases are to be forecast for places where initially none of the harmful organisms is present. Ignorance of pest dynamics or of disease epidemiology is one source of error in forecasting; another is ignorance of the relationship between organism numbers and the harm they cause. On the one hand, crop yield may be little affected despite the presence of many pests, for it will depend on the relative timing of the life cycles of both crop and pest. On the other hand, an initial invasion by a few virus-carrying insects or a few fungal spores may lead to large and intense outbreaks, and consequent serious loss of yield. Systematic verification of forecasts, by comparing them with subsequent events, provides an opportunity to improve our understanding of population dynamics and of epidemiology.

Forecasts of pest and disease outbreaks are of value not only for **managing** numbers but also for **monitoring** their changes. Of particular value is the monitoring of first arrivals of harmful organisms. For example, much time and effort can be saved by avoiding the surveying of crops, or the sorting of insect trap catches for a particular pest species, before significant numbers have become airborne. Monitoring subsequent changes is not only essential for efficient management but also a way of providing the information needed for verification of forecasts.

Monitoring of field populations is a necessary part of forecasting. The area covered must be at least as large as that over which the population is mobile. Sometimes that area will be the one occupied by the whole species. In all cases there is a need for a sampling network and a communications system that gathers information centrally, and in time for effective use. The speed at which field data can be gathered and used will vary greatly with the length of the harmful organism's life cycle, and also upon the methods that are available for getting the forecasts to potential users, whether they are individual farmers or large organisations, local, national or international, that have the responsibility for pest and disease control over wide areas.

Monitoring of field populations, their environment and the harm they cause, should be done as frequently as needed, sometimes even daily. It can be done by means of both crop surveys and the trapping of windborne organisms, and at both permanent sites within a network as well as temporary ones. Information may be gathered by individual farmers, by agricultural advisers, by specialists from research institutes, or by casual observers, but in all cases it must be made available quickly enough for forecasts to be issued in time to be useful. Field data can reach the forecaster directly by telephone, telex or radio, or through agricultural advisers or control organisation staff who collate and synthesise the reports. For ease of comparison over large areas, monitoring and reporting methods should be standardised; they should also be easy to use, yet quantitative. Visualisation of the present state is aided by mapping the reports, either by hand or by machine (if reports are in a standardised code).

Forecasts may be issued irregularly, as and when the need arises, or they may be issued regularly through newspapers, radio, TV, mail or automatic telephone services, and again either directly or through agricultural advisers. Individual farmers use forecasts to help fix the timing of field monitoring and, after modification of general forecasts in the light of the results of local monitoring (even field by field), to choose cultural practices, including those that modify organism numbers. Farmers are likely to use forecasts more or less empirically, depending on their ability and willingness to take this advice as well as on their own experience and that of their neighbours. By contrast, agricultural advisers are more likely to be able to use forecasts to calculate more quantitative crop protection advice. Control organisations and plant protection services use forecasts not only for choosing strategies and tactical deployment of staff and materials in advance of expected changes but also for the longer-term allocation of funds and even their requisition from public or other sources, especially in emergencies.

It follows from the above that any forecasting service for a windborne organism should have several basic requirements:

- a system of sampling over the whole area of mobility
- a rapid communications network for collecting field data
- a central analysis and forecasting centre
- a conceptual model that relates pest and disease distribution, and the harm they cause, to changes in host and environment
- a rapid system of distributing forecasts to users.

But such a service can be justified only if its cost does not exceed the cost of the harm that is to be avoided. Some examples of operational forecasting systems for windborne organisms are described in the following sections.

8.2 FORECASTING OUTBREAKS OF WINDBORNE INSECTS

Some pest insects, such as **locusts** and **armyworms**, appear suddenly and frighteningly in large numbers and cause harm more or less straight away. Others, such as **aphids** and **midges**, can arrive unseen in small numbers, and although they may introduce disease organisms they more usually need to build up after one or more generations before they can cause much harm. These differences in behaviour require differences in forecasts and in strategies of monitoring and management. For the first kind, forecasting emphasises the formation and movement of dense populations. Monitoring needs to cover continuously the whole area of movement, and control relies heavily on killing all large populations, especially the dense ones, wherever they occur, often far from the crops at risk. For the second kind, forecasting emphasises the appearance of first arrivals. Monitoring is often only within the crops at risk, and control relies heavily on preventive cultural strategies or on the killing of populations within the crop before they become too dense.

8.2.1 The desert locust

The nature and causes of build-up and redistribution of this highly mobile species have been extensively studied and the results have been outlined on page 158 (solitary-living) and page 170 (swarms). A system of centralised analysis was set up in 1929 because the renewed plague that had started in 1926 drew attention to the lack of understanding of movement by the species. By 1943, sufficient had been learnt about seasonal redistribution that forecasts could be issued monthly by what later became the Anti-Locust Research Centre (now the Centre for Overseas Pest Research), in London. By 1979, the centralised forecasting service had been transferred to FAO headquarters, in Rome. A detailed discussion of the principles of forecasting is given in Pedgley (1981); here it is sufficient to consider an outline.

Field reports of all stages in the life cycle come from both casual observers and trained staff, and from the vast area shown in Fig. 113. During times of plague, when there are many swarms, monitoring depends heavily on casual observers, such as farmers, nomads, government employees, military units, truck drivers and travellers, who are familiar with swarms and who report verbally to police and others who have access to radio or telephone. In turn, the reports are passed to national or regional locust control organisations and then to the forecaster. During times of recession, when there are few or no swarms, monitoring of the much less easily seen low-density populations depends heavily on trained staff from control organisations or research institutes. The forecasting model is based on the very extensive knowledge that has been built up of the biology, ecology and biogeography of the species, and particularly on the effects of rain and temperature on population growth, and of wind and temperature on mobility. Because of intermittency of contact with field populations, continuity in assessment of known and inferred developments is essential. Much use is made of analogues taken from historical records, and of weather maps for interpreting particular recent events. Forecasts are prepared about mid month for the following calendar month or two, thus allowing about 10 days for them to reach users by air mail in the some 60 countries of the locust's invasion area. Each forecast is accompanied by a summary based on the previous month's summary, on reports received since then, and on interpretation of likely events in areas from which no reports have been received. A monthly interval is adequate because the shortest generation time is about two months.

8.2.2 The African armyworm

This species, too, is highly mobile (page 158). Following the very serious attacks in East Africa during 1961, a research programme was set up in the following year to develop control methods. It was found that control was nearly always too late because farmers did not discover the caterpillars until much damage was being done. It was clear that a forecasting service would help both monitoring and control. Studies of the accumulating records of infestations showed that seasonal changes in distribution, and their dependence on weather, made forecasting possible. Aided by the starting up of a network of light traps to monitor moth populations, the first attempt at forecasting was made in the 1969-70 armyworm season by the former East African Agriculture and Forestry Research Organisation. Forecasts were issued on a limited scale by post to agricultural staff. A full-scale service was started in the following season, when forecasts began to be issued more widely, and made available to the public through radio and the daily press. Successful trials with pheromone traps in 1975 led to the starting up of a network of traps to complement the light traps. In 1977, responsibility for the preparation of forecasts passed to the Desert Locust Control Organisation for Eastern Africa. This service has been described by Betts (1976), Odiyo (1979) and Pedgley & Betts (1980); only an outline is given here. Field reports are of two kinds: caterpillar outbreaks, and moth catches in traps. Outbreak reports come mostly from casual observers, especially farmers, usually through agricultural officers, whereas moth reports come from a network of light and pheromone traps manned by trained staff from the countries concerned. Nightly trap catches are reported weekly to the forecaster by telegram. Forecasting is based upon analogues from historical records, using relationships between trap catches and outbreaks in the following 2-4 weeks and within 200 km. Daily weather maps are used to give guidance on the location of wind convergence that is associated with the concentrated breeding needed to produce outbreaks. Forecasts are issued mid week for the following week, thereby allowing a few days for them to be sent to users. Agricultural officers interpret the forecasts for use by farmers, who also receive them by national radio. A weekly interval is used because the generation time can be as short as one month, and caterpillar outbreaks are not usually reported until they are within about a week of pupation, and therefore about three weeks from egg laying by the newly-emerged adults.

8.2.3 Aphids
Many aphid species are serious pests of annual crops. Because their summer hosts are seasonal, these species must fly to and from overwintering hosts in autumn and spring. The fractions surviving these moves may be very small, but these species have the ability to build up numbers rapidly on the summer host. Understanding their windborne movements is helpful not so much in forecasting the *numbers* arriving as in giving warning of the *time* of arrival (Taylor 1973). In principle, timing can be forecast if source populations are monitored, and if the weather suitable for take-off can be forecast. In practice, it is easier and simpler to monitor the aphids once they have become windborne. In Britain, this is done by means of a network of suction traps, the daily catches from which are sorted centrally for the presence of more than thirty economically important species. In this way, weekly warnings of the recent appearance and cross-country spread of these species are sent to farmers and agricultural advisers in time for

action to be taken, including the starting up of in-crop monitoring, for even though aphids may be airborne they may not land and stay in a particular field, or they may stay but be slow to build up.

One example is the **black bean aphid**, *Aphis fabae*. Forecasts of population density on spring-sown field beans, *Vicia faba*, for eighteen areas in southern England have been made since 1970, based on estimates of population sizes at four stages in the life cycle: during autumn flight to overwintering hosts (spindle trees, *Euonymus europaeus*); as overwintering eggs; as new generations on spindle; and during spring flight to beans (Way *et al.* 1981). Forecasts have been shown to become progressively more accurate by using later estimates from the life cycle; using all four estimates gives excellent long-term warning of the extent of control needed, and also short-term warning of the timing of control. But the possibility must be kept in mind that spring flight might take place from outside the area of the trap network.

8.3 FORECASTING OUTBREAKS OF WINDBORNE DISEASES
Compared with insects, little work has yet been done on forecasting the time of arrival of windborne disease organisms, or the numbers involved. Those organisms carried by insects, such as viruses of both plants and animals, should be forecastable by methods similar to those described in section 8.2; those that are carried directly are perhaps more difficult to forecast. Two examples will illustrate the kinds of methods that have been used so far.

8.3.1 Foot-and-mouth virus
There is strong evidence, from both field and laboratory, that this virus can be spread on the wind (page 87). Pigs are more prolific sources than sheep or cattle, but cattle are more likely to become infected because they breathe at a greater rate and are therefore more likely to take up an infective dose. By knowing how source strength varies with time, and by assuming a simple diffusion model for the plume of windborne virus particles, it is possible to calculate aerial concentrations from a knowledge of topography and the changing weather (Gloster *et al.* 1981, Blackall & Gloster 1981). The method had already been successfully tested using historical records of two British outbreaks before its operational use was forcefully demonstrated by correctly forecasting the 1981 outbreak in southern England, despite virus having to be carried across about 100 km of open sea from a source in northern France.

8.3.2 Hay fever
Forecasting the daily aerial pollen concentration involves forecasting the rate of development of sources as well as the take-off and subsequent windborne movement and dispersion of pollen. Operationally, such forecasting must depend on routinely available weather data, rather than more preferable measurements of intensity and depth of turbulence. The same is true when forecasting the

concentration and spread of atmospheric pollution. In studies of pollen concentration over London, Davies & Smith (1973) found that it was reduced by rain that fell on the day of measurement, but not by rain on the previous day or night. After adjusting for seasonal trends, it was found that concentration increased with daytime maximum temperature and duration of bright sunshine, presumably because these dominated the rates of development of pollen. It was also found that concentration decreased in stronger winds and deeper convection (as demonstrated by the presence of mid-day large convective clouds). Using the previous day's measurement together with forecast changes in the weather, it is possible to forecast changes in pollen concentration. In the Netherlands, a system of daily forecasting of hay fever has been developed based on empirically found relationships between weather and the subjective reactions of about 50 sufferers (Spieksma 1980). A third of the errors in the hay fever forecasts were due to errors in the weather forecasts.

For the hay-fever sufferer, such forecasts are more valuable than statements of current concentration because they give greater warning, and they can also give some idea of the changes in concentration with space and time (Dingle 1955).

References

Adams, A. N. (1967). The vectors and alternative hosts of groundnut rosette virus in Central Province, Malawi, *Rhod. Zamb. Mal. J. agric. Res.,* **5,** 145-151.

Adams, A. P. & Spendlove, J. C. (1970). Coliform aerosols emitted by sewage treatment plants, *Science,* **169,** 1218-1220.

Aidley, D. J. & Lubega, M. (1979). Variation in wing length of the African armyworm, *Spodoptera exempta,* in East Africa during 1973-74, *J. appl. Ecol.,* **16,** 653-662.

Alberts, S. A., Kennedy, M. K. & Cardé, R. T. (1981). Pheromone-mediated anemotactic flight and mating behaviour of the sciarid fly *Bradysia impatiens, Env. Entomol.,* **10,** 10-15.

Andrews, K. L., Barnes, M. M. & Josserand, S. A. (1980). Dispersal and oviposition by navel orangeworm moths, *Env. Entomol.,* **9,** 525-529.

Anon. (1981). Test on the releasing and recapturing or marked planthoppers, *Nilaparvata lugens* and *Sogatella furcifera. Acta ecol. Sinica,* **1,** 49-53.

Arbogast, R. T. (1966). Migration of *Agraulis vanillae* (Lep.: Nymphalidae) in Florida. *Fla. Entomol.,* **49,** 141-145.

Arsdel, E. P. van (1965). Relationships between night breezes and blister rust spread on Lake States white pines, *U.S. Forest Service, Res. Note LS-60.*

Arthur, A. P. & Bauer, D. J. (1981). Evidence of the northerly dispersal of the sunflower moth by warm winds, *Env. Entomol.,* **10,** 528-533.

Austin, B. (1968). Effects of airspeed and humidity changes on spore discharge in *Sordaria fimicola, Ann. Bot.,* **32,** 251-260.

Avitabile, A., Morse, R. A. & Boch, R. (1975). Swarming honeybees guided by pheromones, *Ann. entomol. Soc. Amer.,* **68,** 1079-1082.

Aylor, D. E. (1975a). Force required to detach conidia of *Helminthosporium maydis, Plant Physiol,* **55,** 99-101.

Aylor, D. E. (1975b). Deposition of particles in a plant canopy, *J. appl. Meteorol.,* **14,** 52-57.

212 References

Aylor, D. E. (1976). Resuspension of particles from plant surfaces by wind. In (Eds.) Engelmann, R. J. & Sehmel, G. A. *Atmosphere-surface exchange of particulate and gaseous pollutants.* U.S. Dept. Commerce, Springfield, Va.

Aylor, D. E. (1978). Dispersal in time and space: aerial pathogens. In (Eds.) Horsfall, J. G. & Cowling, E. B. *Plant disease: an advanced treatise.* II, Ch. 8.

Aylor, D. E. & Lukens, R. J. (1974). Liberation of *Helminthosporium maydis* spores by wind in the field, *Phytopathology*, **64**, 1136-1138.

Bainbridge, A. & Legg, B. J. (1976). Release of barley mildew conidia from shaken leaves, *Trans Br. mycol. Soc.*, **66**, 495-498.

Baker R. (1978). *The evolutionary ecology of animal navigation*, Hodder & Stoughton, London.

Baker, T. C. & Roelofs, W. L. (1981). Initiation and termination of oriental fruit moth male response to pheromone concentrations in the field, *Env. Entomol.*, **10**, 211-218.

Baltensweiler, W. & von Salis, G. (1975). Zur dispersionsdynamik der Falter des Grauen Lärchenwicklers (*Zeiraphera diniana*, Tortricidae), *Z. ang. Entomol.*, **77**, 251-257.

Baltensweiler, W. & Fischlin, A. (1979). The role of migration for the population dynamics of the larch bud moth, *Zeiraphera diniana* (Lep.: Tortricidae), *Mitt. Schweiz ent. Ges.*, **52**, 259-271.

Bariola, L. A., Keller, J. C., Turley, D. L. & Farris, J. R. (1973). Migration and population studies of the pink bollworm in the arid west, *Env. Entomol.*, **2**, 205-208.

Basedow, T. (1977). Uber den Flug der Weizengallmücken *Contarinia tritici* und *Sitodiplosis mosellana* in Beziehung zur Windrichtung und zu Weizenfeldern, *Z. ang. Entomol.*, **83**, 173-183.

Batten, A. (1967). Seasonal movements of swarms of *Locusta migratoria migratorioides* in western Africa in 1928 to 1931, *Bll. ent. Res.*, **57**, 357-380.

Batten, A. (1972). Early stages of the 1928-1941 plague of the African migratory locust, *Locusta migratoria migratorioides, Proc. Int. Study Conf. Current and Future Problems of Acridology (London 1970)*, 331-334.

Batzer, H. O. (1968). Hibernation site and dispersal of spruce budworm larvae as related to damage of sapling balsam fir, *J. econ. Entomol.*, **61**, 216-220.

Baumhover, A. H. (1966). Eradication of the screw-worm fly, *J. Amer. med. Assocn.*, **196**, 240-248.

Bauske, R. J. (1967). Dissemination of waterborne *Erwinia amylovora* by wind in nursery plantings, *Proc. Amer. Soc. Hort. Sci.*, **91**, 795-801.

Bauske, R. J. (1968). Bacterial strands: a possible role in fire blight, *Iowa State J. Sci.*, **43**, 119-124.

Bauske, R. J. (1971). Wind dissemination of waterborne *Erwinia amylovora* form *Pyrus* to *Pyracantha* and *Cotoneaster*, *Phytopathology*, **61**, 741-742.

Baust, J. G., Benton, A. H. & Aumann, G. D. (1981). The influence of off-shore platforms on insect dispersal and migration, *Bull. ent. Soc. Amer.*, **27**, 23-25.

Beckwith, R. C. (1978). Biology of the insect. In (Eds.) Brookes, M. H., Stark, R. W. & Campbell, R. W. *The Douglas-fir tussock moth: a synthesis*, Ch. 2.

Beer, S. V. (1979). Fireblight inoculum: sources and dissemination, *EPPO Bull.*, **9**, 13-25.

Berryman, A. A. (1978). Population cycles of the Douglas-fir tussock moth (Lep.: Lymantriidae): the time-delay hypothesis, *Can Entomol.*, **110**, 513-518.

Bertrand, R. F. & English, H. (1976). Release and dispersal of conidia and ascospores of *Valsa leucostoma*. *Phytopathology*, **66**, 987-991.

Bess, H. A. (1974). *Hedylepta blackburni*, a perennial pest of coconut on wind-swept sites in Hawaii, *Proc. Hawaiian ent. Soc.*, **21**, 343-353.

Betts, E. (1976). Forecasting infestations of tropical migrant pests: the desert locus and the African armyworm. In (Ed.) Rainey, R. C. *Insect flight*, 7th symposium, Roy. ent. Soc., London: 113-134.

Billing, E. (1980). Fireblight in Kent, England in relation to weather (1955-1976), *Ann. appl. Biol.*, **95**, 341-364.

Blackall, R. M. & Gloster, J. (1981). Forecasting the airborne spread of foot an mouth disease, *Weather*, **36**, 162-167.

Blair, B. W., Rose, D. J. W. & Law, A. R. (1980). Synoptic weather associated with outbreaks of African armyworm, *Spodoptera exempta* (Lep.: Noctuidae) in Zimbabwe during 1973 and 1976/77, *Zimbabwe J. argic. Res.*, **18**, 95-110.

Blanchard, D. C. & Syzdek, L. (1970). Mechanism for the water-to-air transfer and concentration of bacteria, *Science*, **170**, 626-628.

Böcher, J. (1975). Dispersal of *Nysius groenlandicus* (Het.: Lygaediae) in Greenland, *J. entomol. Medd.*, **43**, 105-109.

Bohaychuck, W. P. & Whitney, R. D. (1973). Environmental factors influencing basidiospore discharge in *Polyporus tomentosus*, *Can. J. Bot.*, **51**, 801-815.

Bowden, J. (1973a). Migration of pests in the tropics, *Med. Fak. Landbouw.*, **38**, 785-795.

Bowden, J. (1973b). The influence of moonlight on catches of insects in light-traps in Africa; I. The moon and moonlight, *Bull. ent. Res.*, **63**, 113-128.

Bowden, J., Brown, G. & Stride, T. (1979). The application of X-ray spectrometry to analysis of migration of *Noctua pronuba*, *Ecol. Entomol.*, **4**, 199-204.

Bowden, J. & Church, B. M. (1973). The influence of moonlight on catches of insects in light traps in Africa; II. The effect of moonphase on light-trap catches, *Bull. ent. Res.*, **63**, 129-142.

Bowden, J. & Gibbs, D. G. (1973). Light-trap and suction-trap catches of insects in the northern Gezira, Sudan, in the season of southward movement of the inter-tropical front, *Bull. ent. Res.*, **62**, 571-596.

Bowden, J., Gregory, P. H. & Johnson, C. G. (1971). Possible wind transport of coffee-leaf rust across the Atlantic Ocean, *Nature*, **229**, 500-501.

Bowden, J. & Morris, M. G. (1975). The influence of moonlight on catches of insects in light-traps in Africa; III. The effective radius of a mercury-vapour light-trap and the analysis of catches using effective radius, *Bull. ent. Res.*, **65**, 303-348.

Boyle, W. W. (1957). On the mode of dissemination of the two-spotted spider mite, *Tetranychus telarius*, *Proc. Hawaiian ent. Soc.*, **16**, 261-268.

Brantjes, N. B. M. (1981). Wind as a factor influencing flower-visiting by *Hadena bicruris* and *Deilephila elpenor*, *Ecol. Entomol.*, **6**, 361-363.

Brodie, H. J. (1951). The splash-cup dispersal mechanism in plants, *Can. J. Bot.*, **29**, 224-234.

Brodie, H. J. (1955). Springboard plant dispersal mechanisms operated by rain, *Can., J. Bot.*, **33**, 156-167.

Brodie, H. J. (1957). Raindrops as plant dispersal agents, *Proc. Indiana Acad. Sci.*, **66**, 65-73.

Brown, C. E. (1958). Dispersal of the pine needle scale, *Phenacaspis pinifoliae*, *Can Entomol.*, **90**, 685-690.

Brown, C. E. (1962). The life history and dispersal of the Bruce spanworm, *Operophtera bruceata* (Lep.: Geometridae), *Can. Entomol.*, **94**, 1103-1107.

Brown, C. E. (1965). Mass transport of forest tent caterpillar moths, *Malacosoma disstria*, by a cold front, *Can. Entomol.*, **97**, 1073-1075.

Brown, E. S. (1970). Nocturnal insect flight direction in relation to wind, *Proc. R. ent. Soc.*, **A45**, 39-43.

Brown, E. S., Betts, E. & Rainey, R. C. (1969). Seasonal changes in distribution of the African armyworm, *Spodoptera exempta*, with special reference to eastern Africa, *Bull. ent. Res.*, **58**, 661-728.

Brown, E. S. & Swaine, G. (1966). New evidence on the migration of the African armyworm, *Spodoptera exempta* (Lep.: Noctuidae), *Bull. ent. Res.*, **56**, 671-684.

Brown, J. S., Kellock, A. W. & Paddick, R. G. (1978). Distribution and dissemination of *Mycosphaerella graminicola* in relation to the epidemiology of speckled leaf blotch of wheat, *Australian J. agric. Res.*, **29**, 1139-1145.

Browning, K. A. (1981). Ingestion of insects by intense convective updraughts, *Antenna*, **5**, 14-16.

Brust, R. A. (1980). Dispersal behaviour of adult *Aedes sticticus* and *Aedes vexans* in Manitoba, *Can. Entomol.*, **112**, 31-45.

Buschman, L. L., Pitre, H. N., Hovermale, C. H. & Edwards, N. C. (1981). Occurrence of the velvetbean caterpillar in Mississippi: winter survival or immigration, *Env. Entomol.*, **10**, 45-52.

Butler, T., Peterson, J. E. & Corbet, P. S. (1975). An exceptionally early and informative arrival of adult *Anax junius* in Ontario, *Can. Entomol.*, **107**, 1253-1254.

Bytinski-Salz, H. (1966). Observations on migrating moths, *Israel J. Entomol.*, **1**, 193.

Callahan, P. S., Sparks, A. N., Snow, J. W. & Copeland, W. W. (1972). Corn earworm moth: vertical distribution in nocturnal flight, *Env. Entomol.*, **1**, 497-503.

Cameron, E. A., McManus, M. L. & Mason, C. J. (1979). Dispersal and its impact on the population dynamics of the gypsy moth in the USA, *Mitt. Schweiz ent. Ges.*, **52**, 169-179.

Campbell, C. A. M. (1977). Distribution of damson-hop aphid (*Phorodon humuli*) migrants on hops in relation to hop variety and wind shelter, *Ann. appl. Biol.*, **87**, 315-325.

Campbell, M. M. & Kettle, D. S. (1979). Swarming of *Culicoides brevitarsis* with reference to markers, swarm size, proximity of cattle, and weather, *Australian J. Zool.*, **27**, 17-30.

Campion, D. G., Bettany, B. W., McGinnigle, J. B. & Taylor, L. R. (1977). The distribution and migration of *Spodoptera littoralis* (Lep.: Noctuidae) in relation to meteorology on Cyprus, interpreted from maps of pheromone trap samples, *Bull. ent. Res.*, **67**, 501-522.

Campion, D. G., Bettany, B. W. & Steedman, R. A. (1974). The arrival of male moths of the cotton leafworm, *Spodoptera littoralis* (Lep.: Noctuidae) at a new continuously recording pheromone trap, *Bull. ent. Res.*, **64**, 379-386.

Cardé, R. T. (1976). Utilization of pheromones in the population management of moth pests, *Env. Health Perspectives*, **14**, 133-144.

Cardé, R. T. & Hagaman, T. E. (1979). Behavioural responses of the gypsy moth in a wind tunnel to airborne enantiomers of disparlure, *Env. Entomol.*, **8**, 475-484.

Carnegie, S. F. (1980). Aerial dispersal of the potato gangrene pathogen, *Phoma exigua*, var. *foveata*, *Ann. appl. Biol.*, **94**, 165-173.

Carter, M. V. (1965). Ascospore deposition in *Eutypa armeniacae*, *Australian J. agric. Res.*, **16**, 825-836.

Casimir, M. & Bament, R. C. (1974). An outbreak of the Australian plague locust, *Chortoicetes terminifera* during 1966-67 and the influence of weather on swarm flight, *Anti-Locust Research Centre*, London, Mem. 12.

Casimir, M. & Edge, V. E. (1979). The development and impact of a control campaign against *Austacris guttulosa* in New South Wales, *PANS*, **25**, 223-236.

Chalfant, R. B., Creighton, C. S., Greene, G. L., Mitchell, E. R., Stanley, J. M. & Webb, J. C. (1974). Cabbage looper: populations in BL traps baited with sex pheromone in Florida, Georgia and South Carolina, *J. Econ. Entomol.*, **67**, 741-745.

Chalmers-Hunt, J. M. (1977). The 1976 invasion of the Camberwell beauty (*Nymphalis antiopa*), *Entomol. Rec. J. Var.*, **99**, 89-105.

Chamberlain, A. C. (1967). Deposition of particles to natural surfaces. In (Eds.) Gregory, P. H. & Monteith, J. L. *Airborne microbes*. C.U.P., Cambridge.

Chang, S-S., Lo, Z-C., & Keng, C-G. (1980). Studies on the migration of rice leaf roller *Cnaphalocrocis medinalis, Acta entomol. Sinica*, **23**, 130-140.

Chapman, J. A. (1962). Field studies on attack flight and log selection by the ambrosia beetle *Trypodendron lineatum* (Col.: Scolytidae), *Can. Entomol.*, **94**, 74-92.

Chapman, J. A. (1967). Response behaviour of scolytid beetles and odour meteorology, *Can. Entomol.*, **99**, 1132-1137.

Chapman, R. F. (1959). Observations on the flight activity of the red locust *Nomadacris septemfasciata, Behaviour*, **14**, 300-334.

Cheng, S-N., Chen, J-C., Si, H., Yan, L-M., Chu, T-L., Wu, C-T., Chien, J-K. & Yan, C-S. (1979). Studies on the migrations of brown planthopper *Nilaparvata lugens, Acta entomol. Sinica*, **22**, 1-20.

Chiykowski, L. N. & Chapman, R. H. (1965). Migration of the six-spotted leafhopper in central North America, *Univ. Wisconsin, Agric. Expl. Stn., Res. Bull.*, **261**, 21-45.

Choudhury, J. H. & Kennedy, J. S. (1980). Light versus pheromone-bearing wind in the control of flight direction by bark beetles, *Scolytus multistriatus, Physiol. Entomol.*, **5**, 207-214.

Christie, A. D. & Ritchie, J. C. (1969). On the use of isentropic trajectories in the study of pollen transports, *Naturaliste Can.*, **96**, 531-549.

Claflin, L. E., Stuteville, D. L. & Armbrust, D. V. (1973). Wind-blown soil in the epidemiology of bacterial leaf spot of alfalfa and common blight of bean, *Phytopathology*, **63**, 1417-1419.

Clark, D. P. (1969). Night flights of the Australian plague locust, *Chortoicetes terminifera*, in relation to storms, *Australian J. Zool.*, **17**, 329-352.

Close, R. C. & Tomlinson, A. I. (1975). Dispersal of the grain aphid *Macrosiphum miscanthi* from Australia to New Zealand, *N. Z. Entomol.*, **6**, 62-65.

Cochrane, J. (1980). Meteorological aspects of the numbers and distribution of the rose-grain aphid, *Metopolophium dirhodum*, over south-east England in July 1979, *Plant Pathol.*, **29**, 1-8.

Cole, F. W. (1975). *Introduction to meteorology*, John Wiley, New York.

Common, I. F. B. (1954). A study of the ecology of adult bogong moth, *Agrotis infusa* (Lep.: Noctuidae), with special reference to its behaviour during migration and aestivation, *Australian J. Zool.*, **2**, 223-263.

Common, I. F. B. (1958). The Australian cutworms of the genus *Agrotis* (Lep.: Noctuidae), *Australian J. Zool.*, **6**, 70-87.

Cooper, J. I. & Harrison, B. D. (1973). The role of weed hosts and the distribution and activity of vector nematodes in the ecology of tobacco rattle virus, *Ann. appl. Biol.*, **73**, 53-66.

Corbet, P. S. (1979). *Pantala flavescens* in New Zealand, *Odontologica*, **8**, 115-121.

Cornwell, P. B. (1960). Movement of the vectors of virus diseases of cacao in Ghana, II — Wind movements and aerial dispersal, *Bull. ent Res.*, **51**, 175-201.

Cour, P., Seignalet, C., Guérin, B., Mayrand, L., Nilsson, S. & Michel, P. B. (1980). In (Ed.) Federal Environmental Agency (of West Germany). *Proc. 1st Internat. Conf. Aerobiol., Munich 1978*, 61-80.

Davey, J. T. (1959). The ecology of *Locusta* in the semi-arid lands and seasonal movements of populations, *Locusta* No. 7, (Bull. Internat. Afr. Mig. Loc. Org.).

Davies, R. R. & Smith, L. P. (1973). Weather and the grass pollen content of the air, *Clinical Allergy*, **4**, 95-108.

Davis, R. (1964). Autecological studies of *Rhynacus breitlowi*, *Fla. Entomol.*, **47**, 113-121.

Daykin, P. N., Kellogg, F. E. & Wright, R. H. (1965). Host-finding and repulsion of *Aedes aegypti*, *Can. Entomol.*, **97**, 239-263.

Den Boer, M. H. (1978). Isoenzymes and migration in the African armyworm, *Spodoptera exempta* (Lep.: Noctuidae), *J. Zool., Lond.*, **185**, 539-553.

Dickson, R. C. & Laird, E. F. (1967). Fall dispersal of green peach aphids to desert valleys, *Ann. entomol. Soc. Amer.*, **60**, 1088-1091.

Digby, P. S. B. (1958). Flight activity in the blowfly *Calliphora erythrocephalus* in relation to wind speed, with special reference to adaptation, *J. expl. Biol.*, **35**, 776-795.

Dindonis, L. L. & Miller, J. R. (1981). Host-finding behaviour of onion flies, *Hylemya antiqua*, *Env. Entomol.*, **9**, 769-772.

Dingle, A. N. (1955). A meteorologic approach to the hay fever problem, *J. Allergy*, **26**, 197-304.

Donaldson, A. I. (1972). The influence of relative humidity on the aerosol stability of different strains of foot-and-mouth disease virus suspended in saliva, *J. gen. Virol.*, **15**, 25-33.

Donaldson, A. I. & Ferris, N. P. (1975). The survival of foot-and-mouth disease virus in open air conditions, *J. Hygiene (Camb.)*, **74**, 409-416.

Donn, W. L. (1975). *Meteorology*, McGraw Hill, New York.

Douthwaite, R. J. (1978). Some effects of weather and moonlight on light-trap catches of the armyworm, *Spodoptera exempta*, at Muguga, Kenya, *Bull. ent. Res.*, **68**, 533-542.

Dowding, P. (1969). The dispersal and survival of spores of fungi causing blue-stain in pines, *Trans. Br. mycol. Soc.*, **52**, 125-137.

Downes, J. A. (1955). Observations on the swarming flight and mating of *Culicoides* (Dipt.: Ceratopogonidae), *Trans. R. entomol. Soc.*, **106**, 213-236.

Downes, J. A. (1969). The swarming and mating flight of Diptera, *Ann. Rev. Entomol.*, **14**, 271-298.

Drake, D. C. & Chapman, R. H. (1965). Evidence for long-distance migration of the six-spotted leafhopper into Wisconsin, *Univ. Wisconsin, Agric. Expl. Stn., Res. Bull.* 261, 3-20.

Drake, V. A., Helm, K. F., Readshaw, J. L. & Reid, D. G. (1981). Insect migration across Bass Strait during spring: a radar study, *Bull. ent. Res.*, **71**, 449-466.

Draper, J. (1980). The direction of desert locust migration, *J. anim. Ecol.*, **49**, 959-974.

Druett, H. A. (1973). The open air factor, *Proc. 4th Internat. Symp. Aerobiol.*, 141-149.

Duelli, P. (1980). Adaptive dispersal and appetitive flight in the green lacewing, *Chrysopa carnea, Ecol. Entomol.*, **5**, 213-220.

Dumont, H. J. (1976). On migration of *Hemianax ephippiger* and *Tramea basilaris* in west and north-west Africa in the winter of 1975/1976, *Odonatologia*, **6**, 13-17.

Dunn, E. (1949). Colorado beetle in the Channel Islands 1947 and 1948, *Ann. appl. Biol.*, **36**, 525-534.

Duviard, D. (1977). Migrations of *Dysdercus* spp. (Hem.: Pyrrhocoridae) related to movements of the inter-tropical convergence zone in West Africa, *Bull. ent. Res.*, **67**, 185-204.

Ebbels, D. L. & King, J. E. (1979) (Eds.) *Plant health,* Blackwell, Oxford.

Edmonds, R. L. (1980). Airborne dispersal of Douglas-fir tussock moth larvae. In (Ed.) Federal Environmental Agency (of West Germany), *Proc. 1st Internat. Conf. Aerobiol.*, Munich 1978, 201-211.

Edroma, E. L. (1977). Outbreak of the African armyworm (*Spodoptera exempta*) in Rwenzori National Park, Uganda, *East Afr. Wildlife*, **15**, 157-158.

Edwards, D. K. (1961). Activity of two species of *Calliphora* (Diptera) during barometric pressure changes of natural magnitude, *Can. J. Zool.*, **39**, 623-635.

Ehrlich, R., Miller, S. & Walker, R. L. (1970). Effects of atmospheric humidity and temperature on the survival of airborne *Flavobacterium, Appl. Microbiol.*, **20**, 884-887.

Erdtman, G. (1937). Pollen grains recovered from the atmosphere over the Atlantic, *Acta Hort. Gothoburg*, **12**, 185-196.

Farkas, S. R. &, Shorey, H. H. (1974). In (Ed.) Birch, M. C. *Pheromones.* Elsevier, Ch. 5.

Farrow, R. A. (1974). Comparative plague dynamics of tropical *Locusta, Bull. ent. Res.*, **64**, 401-411.

Farrow, R. A. (1977). Origin and decline in the 1973 plague locust outbreak in central Western New South Wales, *Australian J. Zool.*, **25**, 455-489.

Farrow, R. A. (1979). Population dynamics of the Australian plague locust, *Chortoicetes terminifera*, in central western New South Wales, I. Reproduction and migration in relation to weather, *Australian J. Zool.*, **27**, 717-745.

Faure, J. C. (1943). Phase variation in the armyworm *Laphygma exempta*, *S. Afr. Dept. Agric. For., Sci. Bull.* 234.

Finch, S. & Skinner, G. (1975). Dispersal of the cabbage root fly, *Ann. appl. Biol.*, **81**, 1-19.

Fleschner, C. A., Badgley, M. E., Ricker, D. W. & Hall, J. C. (1956). Air drift of spider mites, *J. econ. Entomol.*, **49**, 624-627.

Fletcher, B. S. (1973). Observations on a movement of insects at Heron Island, Queensland, *J. Australian Entomol. Soc.*, **12**, 157-160.

Fletcher, J. C. (1978). The African bee, *Apis mellifera adansonii*, in Africa, *Ann. Rev. Entomol.*, **23**, 131-171.

Forster, G. F. (1977). Effect of leaf surface wax on the deposition of airborne propagules, *Trans. Br. mycol. Soc.*, **68**, 245-250.

Fox, K. J. (1970). More records of migrant Lepidoptera in Taranaki and the South Island, *New Zea. Entomol.*, **4**, 63-66.

Fox, K. J. (1978). The transoceanic migration of Lepidoptera to New Zealand — a history and a hypothesis on colonisation, *New Zea. Entomol.*, **6**, 368-380.

Fried, P. M. & Stuteville, D. L. (1977) *Peronospora trifoliorum* sporangium development and effects of humidity and light on discharge and germination, *Phytopathology*, **67**, 890-894.

Frisch, K. von (1967). *The dance language and orientation of bees*, Harvard Univ. Press.

Garms, R., Walsh, J. F. & Davies, J. B. (1979). Studies on the reinvasion of the Onchocerciasis Control Programme in the Volta River basin by *Simulium damnosum* s. 1. with emphasis on the south-western areas, *Tropenmed. Parasit.*, **30**, 345-362.

Garrett-Jones, C. (1962). The possibility of active long-distance migration by *Anopheles pharoensis, Bull. WHO*, **27**, 299-302.

Gatehouse, A. G. & Hackett, D. S. (1980). A technique for studying flight behaviour of tethered *Spodoptera exempta* moths, *Physiol. Entomol.*, **5**, 215-222.

Gatehouse, A. G. & Lewis, C. T. (1973). Host location behaviour of *Stomoxys calcitrans, Entomol. Expl. & appl.*, **16**, 275-290.

Gaylor, M. J. & Frankie, G. W. (1979). The relationship of rainfall to adult flight activity; and of soil moisture to oviposition behaviour and egg and first instar survival in *Phyllophaga crinita, Env. Entomol.*, **8**, 591-594.

Geiger, R. (1965). *The climate near the ground*, 4th edition, Harvard Univ. Press.

Gibo, D. L. (1981a). Altitudes attained by migrating monarch butterflies, *Danaus plexippus* (Lep.: Danaidae), as reported by glider pilots, *Can. J. Zool.*, **59**, 571-572.

Gibo, D. L. (1981b). Some observations on soaring flight in the mourning cloak butterfly (*Nymphalis antiopa*) in southern Ontario, *Trans. New York Entomol. Soc.*, **89**, 98-101.

Gibo, D. L. & Pallett, M. J. (1979). Soaring flight of monarch butterflies, *Danaus plexippus*, during the late summer migration in southern Ontario, *Can. J. Zool.*, **57**, 1393-1401.

Gibson, W. W. & Painter, R. H. (1957). Transportation by aphids of the wheat curl mite, *Aceria tulipae*, a vector of wheat streak mosaic virus, *J. Kansas Entomol. Soc.*, **30**, 147-153.

Gillett, J. D. (1979). Out for blood: flight orientation upwind in the absence of visual clues, *Mosquito News*, **39**, 221-229.

Gillies, M. T. & Wilkés, T. J. (1969). A comparison of the range of attraction of animal baits and of carbon dioxide for some West African mosquitoes, *Bull. ent. Res.*, **59**, 441-456.

Gillies, M. T. & Wilkes, T. J. (1970). The range of attraction of single baits for some West African mosquitoes, *Bull. ent. Res.*, **60**, 225-235.

Gillies, M. T. & Wilkes, T. J. (1972). The range of attraction of animal baits and carbon dioxide for mosquitoes. Studies in a freshwater area of West Africa, *Bull. ent. Res.*, **61**, 389-404.

Gillies, M. T. & Wilkes, T. J. (1981). Field experiments with a wind tunnel on the flight speed of some West African mosquitoes (Dip.: Culicidae), *Bull. ent. Res.*, **71**, 65-70.

Girard, P. J. (1947). The Colorado beetle in Guernsey and the neighbouring islands, 1947, *Rep. Trans. Soc. Guernesiaise*, **14**, 154-161.

Glasscock, H. H. (1971). Fireblight epidemic among Kentish apple orchards in 1969, *Ann. appl. Biol.*, **69**, 137-145.

Glick, P. A. (1965). Review of collections of Lepidoptera by airplane, *J. Lepidopterists Soc.*, **19**, 129-137.

Glick, P. A. (1967). Aerial dispersal of the pink bollworm in the United States and Mexico, *USDA, Agric. Res. Service, Production Res. Rep. No. 96*.

Gloster, J., Blackall, R. M. Sellers, R. F. & Donaldson, A. I. (1981). Forecasting the airborne spread of foot-and-mouth disease, *Vet. Record*, **108**, 370-374.

Goldsmith, A., Chaing, H. C. & Okubo, A. (1980). Turning motion of individual midges, *Anarete pritchardi*, in swarms, *Ann. entomol. Soc. Amer.*, **73**, 526-528.

Grace, J. (1977). *Plant responses to the wind*, Academic, New York.

Graham, D. C. & Harrison, M. D. (1975). Potential spread of *Erwinia* spp. in aerosols, *Phytopathology*, **65**, 739-741.

Graham, D. C., Quinn, C. E. & Bradley, L. F. (1977). Quantitative studies on the generation of aerosols of *Erwinia carotovora* var. *atroseptica* by simulated raindrop impaction on black-leg infected potato stems, *J. appl. Bacteriol.,* **43**, 413-424.

Gransbo, G. (1980). Control measures towards Colarado beetle eradication in Sweden, *EPPO Bull.,* **10**, 499-505.

Gray, B., Billings, R. F., Gara, R. I. & Johnsey, R. L. (1972). On the emergence and initial flight behaviour of the mountain pine beetle, *Dendroctonus ponderosus,* in eastern Washington, *Z. ang. Entomol.,* **21**, 250-259.

Greathead, D. J. (1972). Dispersal of the sugar-cane scale *Aulacaspis tegalensis* (Hem.: Disapididae) by air currents, *Bull. ent. Res.,* **61**, 547-558.

Green, G. W. & Pointing, P. J. (1962). Flight and dispersal of the European pine shoot moth, *Rhyacionia buoliana;* II. Natural dispersal of egg-laden females, *Can. Entomol.,* **94**, 299-314.

Greenbank, D. O., Schaefer, G. W. & Rainey, R. C. (1980). Spruce budworm moth flight and dispersal, *Mem. Entomol. Soc. Can.,* No. 110.

Gregory, P. H. (1949). The operation of the puffball mechanism of *Lycoperdon perlatum* by raindrops shown by ultra-high-speed Schlieren cinematography, *Trans. Br. mycol. Soc.,* **32**, 11-15.

Gregory, P. H. (1971). The leaf as a spore trap. In (Eds.) Preece, T. F. & Dickinson, C. H. *Ecology of leaf surface micro-organisms,* Academic, New York.

Gregory, P. H. (1973). *Microbiology of the atmosphere.* Leonard Hill, London.

Gregory, P. H., Guthrie, E. J. & Bunce, M. E. (1959). Experiments on splash dispersal of fungus spores, *J. gen. Microbiol.,* **20**, 328-354.

Grover, S. N., Pruppacher, H. R. & Hamielec, A. E. (1977). A numerical determination of the efficiency with which spherical aerosol particles collide with spherical water drops due to inertial impaction and phoretic and electrical forces, *J. atmos. Sci.,* **34**, 1655-1663.

Guilmette, J. E., Holzapfel, E. P. & Tsuda, D. M. (1970). Trapping of airborne insects on ships in the Pacific, Part 8, *Pacific Insects,* **12**, 303-325.

Gutierez, A. P., Nix, H. A., Havenstein, D. E. & Moore, P. A. (1974). The ecology of *Aphis craccivora* and subterranean clover stunt virus in south-east Australia; III. A regional perspective of the phenology and migration of the cowpea aphid, *J. appl. Ecol.,* **11**, 21-35.

Gutterman, Y., Witztum, A. & Evenari, M. (1967). Seed dispersal and germination in *Blepharis persica, Israel J. Bot.,* **16**, 213-234.

Haggis, M. J. (1971). Light-trap catches of *Spodoptera exempta* in relation to wind direction, *E. Afr. J. Agric. For.,* **37**, 100-108.

Haggis, M. J. (1981). Spatial and temporal changes in the distribution of eggs of *Heliothis armiger* (Lep.: Noctuidae) on cotton in the Sudan Gezira, *Bull. ent. Res.,* **71**, 183-193.

Haile, D. G., Snow, J. W. & Young, J. R. (1975). Movement by adult *Heliothis* released in St. Croix to other islands, *Env. Entomol.,* **4**, 225-226.

Haine, E. (1955). Aphid take-off in controlled wind speeds, *Nature,* **175,** 474-475.

Hammett, K. R. W. & Manners, J. G. (1974). Conidium liberation in *Erysiphe graminis;* III. Wind tunnel studies, *Trans. Br. mycol. Soc.,* **62,** 267-282.

Handel, S. N. (1976). Restricted pollen flow of two woodland herbs determined by neutron-activation analysis, *Nature,* **260,** 422-423.

Hatch, M. H. & Dimmick, R. L. (1966). Physiological responses of airborne bacteria to shifts in relative humidity, *Bacteriol. Rev.,* **30,** 597-603.

Hawkes, C. (1974). Dispersal of adult cabbage root fly (*Erioischia brassicae*) in relation to a brassica crop, *J. appl. Ecol.,* **11,** 83-93.

Hawkes, C. & Coaker, T. H. (1976). Behavioural responses to host-plant odours in adult cabbage root fly (*Erioischia brassicae*), *Symp. Biol. Hung.,* **16,** 85-89.

Hawkes, C., Patton, S. & Coaker, T. H. (1978). Mechanisms of host-plant finding in adult cabbage root fly, *Delia brassicae, Entomol. exp. & appl.,* **24,** 419-427.

Helm, K. F. (1975). Migration of the armyworm *Persectania ewingii* moths in spring and the origin of outbreaks, *J. Australian entomol. Soc.,* **14,** 229-236.

Hendricks, D. E., Graham, H. M. & Raulston, J. R. (1973). Dispersal of sterile tobacco budworms from release points in northeastern Mexico and southern Texas, *Env. Entomol.,* **2,** 1085-1088.

Hendricks, D. E., Perez, C. T. & Guerra, R. J. (1980). Effects of nocturnal wind on performance of two sex pheromone traps for noctuid moths, *Env. Entomol.,* **9,** 483-485.

Hermansen, J. E. & Stix, E. (1975). Evidence of wind dispersal of powdery mildew conidia across the North Sea, *Roy. Vet. Agric. Univ., Copenhagen, 1974 Yearbook,* 87-100.

Hermansen, J. E., Torp, U. & Prahm, L. (1976). Evidence of distant dispersal of live spores of *Erysiphe graminis* f. sp. *hordei, Roy. Vet. Agric. Univ., Copenhagen, 1975 Yearbook,* 17-30.

Hermansen, J. E. and Wiberg, A. (1972). On the appearance of *Erysiphe graminis* f. sp. *hordei* and *Puccinia hordei* in the Faeroës and the possible primary sources of inoculum, *Friesia,* **10,** 30-34.

Hers, J. F. P. & Winkler, K. C. (Eds.) (1973). *Airborne transmission and airborne infection,* Oosthoek, Utrecht.

Hightower, B. G., Adams, A. L. & Alley, D. A. (1965). Dispersal of released irradiated laboratory-reared screw-worm flies, *J. econ. Entomol.,* **58,** 373-374.

Hill, M. G. (1980). Wind dispersal of the coccid *Icerya seychellarum* (Hom.: Margaroididae) on Aldabra atoll, *J. anim. Ecol.,* **49,** 939-957.

Hirst, J. M. & Stedman, O. J. (1963). Dry liberation of fungal spores by raindrops, *J. gen. Microbiol.,* **33,** 335-344.

Hirst, J. M. & Stedman, O. J. (1971). Patterns of spore dispersal in crops. In (Eds.) Preece, T. F. & Dickinson, C. H. *Ecology of leaf surface micro-organisms,* 91-101.

Hirst, J. M., Stedman, O. E. & Hurst, G. W. (1967). Long-distance spore transport: vertical sections of spore clouds over the sea, *J. gen. Microbiol.,* **48,** 357-377.

Hoelscher, C. E. (1967). Wind dispersal of brown soft scale crawlers, *Coccus hesperidum* (Hom.: Coccidae), and Texas citrus mites, *Eutetranychus banksii* (Acarina: Tetranychidae) from Texas citrus, *Ann. entomol. Soc. Amer.,* **60,** 673-678.

Holsten, E. H. & Gara, R. I. (1975). Flight of the mahogany shoot borer, *Hypsipyla grandella, Ann. entomol. Soc. Amer.,* **68,** 319-320.

Holzapfel, E. P. & Perkins, B. D. (1969). Trapping of airborne insects on ships in the Pacific, Part 7, *Pacific Insects,* **11,** 455-476.

Horsfall, J. G. & Cowling, E. B. (Eds.) (1978). *Plant disease: an advanced treatise,* Vol. II: How disease develops in populations, Academic, New York.

Horsfall, W. R. (1954). A migration of *Aedes vexans, J. econ. Entomol.,* **47,** 544.

Hsia, T-S., Tsai, S-M. & Ten, H-S. (1963). Studies of the regularity of outbreak of the oriental armyworm, *Leucania separata, Acta entomol. Sinica,* **12,** 552-564.

Hugh-Jones, M. E. & Wright, P. B. (1970). Studies on the 1967-8 foot-and-mouth disease epidemic, *J. Hyg., Camb.,* **68,** 253-271.

Hughes, R. D. (1970). The seasonal distribution of bushfly (*Musca velustissima*) in south-east Australia, *J. anim. Ecol.,* **39,** 691-706.

Hughes, R. D. & Nicholas, W. L. (1974). The spring migration of the bushfly (*Musca velustissima*): evidence of displacement provided by natural population markers including parasitism, *J. anim. Ecol.,* **43,** 411-428.

Hurst, G. W. (1969a) *Danaus plexippus* in Britain in 1968, and a possible pattern of immigration from 1933 onwards, *Agric. Mem. No. 307.* (Unpublished report in the library of the British Meteorological Office.)

Hurst, G. W. (1969b). Meteorological aspects of insect migrations, *Endeavour,* **28,** 77-81.

Hurst, G. W. (1969c). Insect migrations to the British Isles, *Quart. J. Roy. meteorol. Soc.,* **95,** 435-439.

Hurst, G. W. (1971). *Danaus plexippus* in Bermuda in September 1970, *Agric. Mem. No. 354.* (Unpublished report in the library of the British Meteorological Office.)

Ingold, C. T. (1971). *Fungal spores: their liberation and dispersal,* Clarendon Press, Oxford.

Inscoe, M. (1977). Chemical communication in insects, *J. Washington Acad. Sci.,* **67,** 16-33.

Jarvis, W. R. (1962). The dispersal of spores of *Botrytis cinerea* in a raspberry plantation, *Trans. Br. mycol. Soc.,* **45,** 549-559.

Jeppson, L. R., Keifer, H. H. & Baker, E. W. (1975). *Mites injurious to economic plants,* Univ. California Press.

Jiang, G-H., Tan, H-Q., Shen, W-Z., Cheng, X-N. & Chen, R-C. (1981). The relation between long-distance northward migration of the brown planthopper (*Nilaparvata lugens*) and synoptic weather conditions, *Acta entomol. Sinica,* **24**, 251-261.

Johnson, B. (1957). Studies on the dispersal by upper winds of *Aphis craccivora* in New South Wales, *Proc. Linn. Soc. N. S. W.,* **82**, 191-198.

Johnson, C. G. (1957). The vertical distribution of aphids in the air and the temperature lapse rate, *Quart. J. Roy. meteorol. Soc.,* **83**, 194-201.

Johnson, C. G. (1969). *Migration and dispersal of insects by flight,* Methuen, London.

Johnson, C. G. & Taylor, L. R. (1957). Periodism and energy summation with special reference to flight rhythms in aphids, *J. exp. Biol.,* **34**, 209-221.

Johnson, D. T. & Croft, B. A. (1976). Laboratory study of the dispersal behaviour of *Amblyseius fallacis* (Acarina: Phytoseiidae), *Ann. ent. Soc. Amer.,* **69**, 1019-1023.

Johnson, D. T. & Croft, B. A. (1981). Dispersal of *Amblyseius fallacis* (Acarina: Phytoseiidae) in an apple ecosystem, *Env. Entomol.,* **10**, 313-319.

Johnson, D. W. & Kuntz, J. E. (1979). *Eutypella* canker of maple: ascospore discharge and dissemination, *Phytopathology,* **69**, 130-135.

Johnson, W. L., Cross, W. H., Leggett, J. E., McGovern, W. L., Mitchell, H. C. & Mitchell, E. B. (1975). Dispersal of marked boll weevil: 1970-1973 studies, *Ann. ent. Soc. Amer.,* **68**, 1018-1022.

Johnson, W. L., Cross, W. H. & McGovern, W. L. (1976). Long-range dispersal of marked boll weevils in Mississippi during 1974, *Ann. ent. Soc. Amer.,* **69**, 421-422.

Jones, D. G. & Clifford, B. C. (1978). *Cereal diseases: their pathology and control,* BASF.

Jones, O. T., Lomer, R. A. & Howse, P. E. (1981). Responses of male Mediterranean fruit flies, *Ceratitis capitata,* to trimedlure in a wind tunnel of novel design, *Physiol. Entomol.,* **6**, 175-181.

Kaae, R. S. & Shorey, H. H. (1973). Sex pheromones of Lepidoptera. 44. Influence of environmental conditions on the location of pheromone communication and mating in *Pectinophora gossypiella, Env. Entomol.,* **2**, 1081-1084.

Kaae, R. S., Shorey, H. H., Gaston, L. K. & Sellers, D. (1977). Sex pheromones of Lepidoptera: seasonal distribution of male *Pectinophora gossypiella* in a cotton growing area, *Env. Entomol.,* **6**, 284-286.

Kainoh, Y., Shimizu, K., Maru, S. & Tamaki, Y. (1980). Host-finding behaviour of the rice bug, *Leptocorisa chinensis* (Hem.: Coreidae), with special reference to diel patterns of aggregation and feeding on rice plant, *Appl. Entomol. Zool.,* **15**, 225-233.

Kanze, J. E. (1977). The orientation of migrant and non-migrant monarch butterflies, *Danaus plexippus, Psyche*, **84**, 120-141.

Katzenelson, E., Buium, I. & Shuval, H. I. (1976). Risk of communicable disease infection associated with wastewater irrigation in agricultural settlements, *Science*, **194**, 944-946.

Kawai, A. (1978). Movement of the sterilized melon fly from Kume Island to adjacent islets, *Appl. Entomol. Zool.*, **13**, 314-315.

Kehat, M., Navon, A. & Greenberg, S. (1976). Captures of marked *Spodoptera littoralis* male moths in virgin female traps: effects of wild male population, distance of traps from release point, and wind, *Phytoparasitica*, **4**, 77-83.

Kellogg, F. E., Frizel, D. E. & Wright, R. H. (1962). The olfactory guidance of flying insects. IV. *Drosophila, Can Entomol.*, **94**, 884-888.

Kelly, C. D. & Pady, S. M. (1953). Microbiological studies of air over some non-arctic regions of Canada, *Can. J. Bot.*, **31**, 90-106.

Kennedy, J. S. (1939). The visual responses of flying mosquitoes, *Proc. zool. Soc.*, **A109**, 221-242.

Kennedy, J. S. (1977). Olfactory responses to distant plants and other odour sources. In (Eds.) Shorey, H. H. & McKelvey, J. J. *Chemical control of insect behaviour*, Ch. 5.

Kennedy, J. S. (1978). The concepts of olfactory 'arrestment' and 'attraction', *Physiol. Entomol.*, **3**, 91-98.

Kennedy, J. S,, Ludlow, A. R. & Sanders, C. J. (1980). Guidance system used in moth sex attraction, *Nature*, **288**, 475-477.

Kennedy, J. S., Ludlow, A. R. & Sanders, C. J. (1981). Guidance of flying male moths by windborne sex pheromone, *Physiol. Entomol.*, **6**, 395-412.

Kennedy, J. S., & Marsh, D. (1974). Pheromone-regulated anemotaxis in flying moths, *Science*, **184**, 999-1001.

Kettlewell, H. B. D. & Heard, M. J. (1961). Accidental radioactive labelling of a migrating moth, *Nature*, **189**, 676-677.

Keyserlingk, H. von (1980). Control of Dutch elm disease by behavioural manipulations of its vectors, *Med. Fac. Landbouww. Rijksuniv. Gent*, **45**, 475-488.

Kieckhefer, R. W., Lytle, W. F. & Spuhler, W. (1974). Spring movement of cereal aphids into South Dakota, *Env. Entomol.*, **3**, 347-350.

Kieckhefer, R. W. & Medler, J. T. (1966). Aggregations of the potato leafhopper in alfalfa fields in Wisconsin, *Ann. ent. Soc. Amer.*, **59**, 180-182.

King, A. M. Q., Underwood, B. O., McCahon, D., Newman, J. W. I. & Brown, F. (1981). Biochemical identification of viruses causing the 1981 outbreaks of foot-and-mouth disease in the UK, *Nature*, **293**, 479-480.

Kirkpatrick, T. W. (1957). *Insect life in the tropics*, Longman & Green, London.

Kisakabe, S. (1979). Dispersal of the brown planthopper, *Nilaparvata lugens*, (Hem.: Delphacidae) in relation to its population growth, *Appl. Entomol. Zool.*, **14**, 224-225.

Kishaba, A. N., Toba, H. H., Wolf, W. W. & Vail, P. V. (1970). Response of laboratory-reared male cabbage looper to synthetic sex pheromone in the field, *J. econ. Entomol.*, **63**, 178-181.

Kisimoto, R. (1976). Synoptic weather conditions inducing long-distance immigration of planthoppers, *Sogatella furcifera* and *Nilaparvata lugens*, *Ecol. Entomol.*, **1**, 95-109.

Kisimoto, R. (1979), Brown planthopper migration. In: *Brown planthopper: threat to rice production in Asia,* (Proc. Internat. Rice Res. Inst. Symp. 1977).

Klassen, W. & Hocking, B. (1964). The influence of a deep river valley system on the dispersal of *Aedes* mosquitoes, *Bull. ent. Res.*, **55**, 289-304.

Knipling, E. F. (1960). The eradication of the screw-worm fly, *Sci. Amer.*, **203**, 54-61.

Koeniger, N. & Koeniger, G. (1980). Observations and experiments on migration and dance communications of *Apis dorsata* in Sri Lanka, *J. apic. Res.*, **19**, 21-34.

Laird, M. (1962). A flight of insects in the Gulf of Aden, *Proc. 11th Internat. Cong. Entomol.*, **3**, 35-36.

Lanier, G. N. & Burns, B. W. (1978). Effects on the responsiveness of bark beetles to aggregation chemicals, *J. Chem. Ecol.*, **4**, 139-147.

Larsen, E. B. (1949). Activity and migration of *Plusia gamma, Biol. Meddr.*, **21(4)**.

Lawson, F. R., Chamberlain, J. C. & York, G. T. (1951). Dissemination of the beet leafhopper in California, *U.S. Dept Agric., Tech. Bull.* 1030.

Lea, A. (1964). Some major factors in the population dynamics of the brown locust, *Locustana pardalina, Monog. Biol.*, **14**, 269-283.

Lea, A. (1969). The distribution and abundance of brown locusts, *Locustana pardalina*, between 1954 and 1965, *J. entomol. Soc. S. Afr.*, **32**, 366-398.

Leach, C. M. (1976). An electrostatic theory to explain violent spore liberation by *Drechslera turcica* and other fungi, *Mycologia*, **68**, 63-86.

Leach, C. M. (1980a). Influence of humidity, red-infrared radiation, and vibration on spore discharge by *Pyricularia oryzae, Phytopathology*, **70**, 201-205.

Leach, C. M. (1980b). Influence of humidity and red-infrared radiation on spore discharge by *Drechslera turcica* — additional evidence, *Phytopathology*, **70**, 192-196.

Leach, C. M. (1980c). Evidence for an electrostatic mechanism in spore discharge by *Drechslera turcica, Phytopathology*, **70**, 206-213.

Le Berre, R., Garms, R., Davies, J. B., Walsh, J. F. & Philippon, B. (1979). Displacements of *Simulium damnosum* and strategy of control against onchocerciasis, *Phil. Trans. Roy. Soc. Lond.*, **B287**, 277-288.

Lecoq, M. (1975). *Les déplacements par vol du criquet migrateur Malgache en phase solitaire. Leur importance sur la dynamique des populations et la grégarisation,* Ministère de Cooperation, Paris.

Lecoq, M. (1978) Les déplacements par vol à grand distance chez les acridiens des zones sahélienne et soudanienne en Afrique de l'ouest, *Compt. Rend. Acad. Sci., Paris,* **286D**, 419-422.

Lee, R. E. (1980). Aggregation of lady beetles on the shores of lakes (Col.: Coccinellidae), *Amer. Midl. Nat.,* **104**, 295-304.

Leonard, D. E. (1971). Airborne dispersal of larvae of the gypsy moth and its influence on concepts of control, *J. econ. Entomol.,* **64**, 638-641.

Lewis, T. (1965a). The effect of an artificial windbreak on the distribution of aphids in a lettuce crop, *Ann. appl. Biol.,* **55**, 513-558.

Lewis, T. (1965b). The effects of an artificial windbreak on the aerial distribution of flying insects, *Ann. appl. Biol.,* **55**, 503-512.

Lewis, T. (1966). Artificial windbreaks and the distribution of turnip mild yellows virus and *Scaptomyza apicalis* (Dip.) in a turnip crop, *Ann. appl. Biol.,* **58**, 371-376.

Lewis, T. (1969a). Factors affecting primary patterns of infestation, *Ann. appl. Biol.,* **63**, 315-317.

Lewis, T. (1969b). The distribution of flying insects near a low hedgerow, *J. appl. Ecol.,* **6**, 443-452.

Lewis, T. (1970). Patterns of distribution of insects near a windbreak of tall trees, *Ann. appl. Biol.,* **65**, 213-220.

Lewis, T. (1973). *Thrips* Academic, New York.

Lewis, T. & Macaulay, E. D. M. (1976). Design and elevation of sex-attractant traps for pea moth, *Cydia nigricana,* and the effect of plume shape on catches, *Ecol. Entomol.,* **1**, 175-187.

Li, K-P., Wong, H-H. & Woo, W-S. (1964). The hypothetical injuriousness of the seasonal migrations of the oriental armyworm and experiments with the release and recapture of marked individuals, *Acta Phytophylac. Sin.,* **3**, 101-110.

Lin, C-S., Sun, C-J., Chen, R-L. & Cheng, J-S. (1963). Studies of the regularity of the outbreak of the oriental armyworm, *Leucania separata.* I. The early spring migration of the oriental armyworm moths and its relation to winds, *Acta. Entomol. Sin.,* **12**, 243-261.

Lingren, P. D., Sparks, A. N., Raulston, J. R. & Wolf, W. W. (1978). Applications for nocturnal studies of insects, *Bull. ent. Soc. Amer.,* **24**, 206-212.

Lloyd, A. B. (1969). Dispersal of Streptomycetes in air, *J. gen. Microbiol.,* **57**, 35-40.

Lokki, J., Malmström. K. K. & Suomalainen, E. (1978). Migration of *Vanessa cardui* and *Plutella xylostella* (Lep.) to Spitzbergen in the summer 1978. *Not. entomol.,* **58**, 121-123.

Ludlam, F. H. (1980). *Clouds and storms.* Pennsylvania State Univ. Press.

Maddison, A. C. & Manners, J. G. (1972). Sunlight and viability of cereal rust uredospores, *Trans. Br. mycol. Soc., 59*, 429-443.

Magor, J. I. & Rosenberg, L. J. (1980). Studies of winds and weather during migrations of *Simulium damnosum* (Dip.: Simuliidae), the vector of onchocerciasis in West Africa, *Bull. ent. Res., 70*, 693-716.

Mahindapala, R. (1978). Epidemiology of maize rust, *Puccinia sorghi, Ann. appl. Biol., 90*, 155-161.

Mankin, R. W., Vick, K. W., Mayer, M. S., Coffelt, J. A. & Callahan, P. S. (1980). Models for dispersal of vapours in open and confined spaces: applications to sex pheromone trapping in a warehouse, *J. chem. Ecol., 6*, 929-950.

Marks, R. J. (1977). The influence of climatic factors on catches of the red bollworm, *Diparopsis castanea* (Lep.: Noctuidae), in sex pheromone traps, *Bull. ent. Res., 67*, 243-248.

Marsh, D., Kennedy, J. S. & Ludlow, A. R. (1978). An analysis of anemotactic zigzagging flight in male moths stimulated by pheromone, *Physiol. Entomol., 3*, 221-240.

Marsh, D., Kennedy, J. S. & Ludlow, A. R. (1981). Analysis of zigzagging flight in moths; a correction, *Physiol. Entomol., 6*, 225.

Martyn, E. J. & Hudson, N. M. (1953). Control of the armyworm *Persectania ewingii* in Tasmania, *Tasmania J. Agric., 34*, 92-94.

Mason, R. R. (1974). Population change in an outbreak of the Douglas-fir tussock moth, *Orgyia pseudotsugata,* in central Arizona, *Can. Entomol., 106*, 1171-1174.

Mayer, M. S. & James, J. D. (1968). Attraction of *Aedes aegyptii*: responses to human arms, carbon dioxide, and air currents in a new type of olfactometer, *Bull. ent. Res., 58*, 629-642.

McClure, M. S. (1977). Dispersal of the scale *Fiorinia externa* (Hom.: Diaspididae) and effects of edaphic factors on its establishment on hemlock, *Env. Entomol., 6*, 539-524.

McClure, M. S. (1979). Spatial and seasonal distribution of disseminating stages of *Fiorinia externa* (Hom.: Diaspididae) and natural enemies in a hemlock forest, *Env. Entomol., 8*, 869-873.

McCutchen, C. W. (1977). The spinning rotation of ash and tulip tree samaras, *Science, 197*, 691-692.

Meade, A. B. & Peterson, A. G. (1964). Origin of populations of the six-spotted leafhopper, *Macrosteles fascifrons,* in Anoka county, Minnesota, *J. econ. Entomol., 57*, 885-888.

Medler, J. T. & Ghosh, A. K. (1968). Apterous aphids in water, wind and suction traps, *J. econ. Entomol., 61*, 267-270.

Melnichenko, A. N. (1936). Regularities of mass flying of the adult *Loxostege sticticalis,* and the problem of the prognosis of their flight migrations, *Bull. Plant Prot., Leningrad,* Ser. 1, No. 17.

Meredith, D. S. (1965). Violnet spore release in *Helminthosporium turcicum*, *Phytopathology*, **55**, 1099-1102.

Meyer, H. J. & Morris, D. M. (1973). A mathematical relation to describe the influence of wind on the initial flight dispersal of *Scolytus multistriatus* (Col.: Scolytidae), *Ann. ent. Soc. Amer.*, **66**, 505-508.

Mikkola, K. (1971). The migratory habit of *Lymantria dispar* (Lep.: Lymantriidae) adults of continental Eurasia in the light of a flight to Finland, *Acta entomol. Fenn.*, **28**, 107-120.

Mikkola, K. (1978). Spring migrations of wasps and bumble bees on the southern coast of Finland, *Ann. entomol. Fenn.*, **44**, 10-26.

Mikkola, K. & Salmensuu, P. (1965). Migration of *Laphygma exigua* (Lep.: Noctuidae) in northwestern Europe in 1964, *Ann. zool. Fenn.*, **2**, 124-139.

Miller, L. W., Martyn, E. J. & Hudson, N. M. (1963). The prediction of outbreaks of southern armyworm, *Persectania ewingii* in Tasmania, *Tasmania J. Agric.*, **34**, 92-94.

Mitchell, E. R. (1979). Migration by *Spodoptera exigua* and *S. frugiperda*, North American style. In (Eds.) Rabb, R. L. & Kennedy, G. G. *Movement of highly mobile insects*, N. Carolina State Univ., 386-393.

Mitchell, E. R., Chalfant, R. B., Greene, G. L. & Creighton, C. S. (1975). Soybean looper: population in Florida, Georgia and South Carolina, as determined with pheromone-baited BL traps, *J. econ. Entomol.*, **68**, 747-750.

Mitchell, R. G. (1979). Dispersal of early instars of the Douglas-fir tussock moth, *Ann. ent. Soc. Amer.*, **72**, 291-296.

Miyahara, Y. & Kawai, A. (1979). Movement of sterilized melon fly from Kume Island to the Amani Islands, *Appl. Entomol. Zool.*, **14**, 496.

Miyahara, Y., Wada, T. & Kobayashi, M. (1981). Appearance of *Cnaphalocrocis medinalis* in early planted rice fields in Chikugo, Japan, *Jap. J. appl. Entomol. Zool.*, **25**, 26-32.

Moar, N. T. (1969a). Possible long-distance transport of pollen to New Zealand, *New Zea. J. Bot.*, **7**, 424-426.

Moar, N. T. (1969b). Pollen analysis of a surface sample from Antipodes Island, *New Zea. J. Bot.*, **7**, 419-423.

Mochida, O. & Takada, H. (1978). Possible migration of aphid parasites across the East China Sea, *Appl. Entomol. Zool.*, **13**, 125-127.

Monteith, J. L. (1973). *The principles of environmental physics.* Arnold, London.

Monteith, J. L. (Ed.) (1975). *Vegetation and the atmosphere, vol. 1.* Academic, London.

Morse, R. A. (1963). Swarm orientation in honey bees, *Science*, **141**, 356-357.

Munro, J. A. & Saugstad, S. (1938). A measure of the flight capacity of grasshoppers, *Science*, **88**, 473-474.

Murlis, J. & Bettany, B. W. (1977). Night flight towards a sex pheromone source by male *Spodoptera littoralis* (Lep.: Noctuidae), *Nature*, **268**, 433-435.

Murlis, J. & Jones, C. D. (1981). Fine-scale structure of odour plumes in relation to insect orientation to distant pheromone and other attractant sources, *Physiol. Entomol.*, **6**, 71-86.

Murray, M. D. (1970). The spread of ephemeral fever of cattle during the 1967-68 epizootic in Australia, *Australian vet. J.*, **46**, 77-82.

Nagarajan, S., Joshi, L. M., Srivastava, K. D. & Singh, D. V. (1980). In (Ed.) Federal Environmental Agency (of West Germany), *Proc. 1st Internat. Conf. Aerobiol.*, *Munich 1978*, 446-451.

Nakamura, K. & Kawasaki, K. (1977). The active space of the *Spodoptera litura* sex pheromone and the pheromone component determining this space, *Appl. Entomol. Zool.*, **12**, 162-177.

Nault, L. R. & Styer, W. E. (1969). The dispersal of *Aceria tulipae* and three other grass-infesting eriophyid mites in Ohio, *Ann. ent. Soc. Amer.*, **62**, 1446-1455.

Nellis, D. W. (1977). Screw-worm fly transmission by wind, *J. Parasitol.*, **63**, 178-179.

Nichiporick, W. (1965). The aerial migration of the six-spotted leafhopper and the spread of the virus disease aster yellows, *Internat. J. Bioclim. Biomet.*, **9**, 219-227.

Nielsen, E. T. (1961). On the habits of the migratory butterfly *Ascia monuste*, *Biol. Meddr.*, **23(11)**.

Nutman, F. J. & Roberts, F. M. (1970). Coffee leaf rust, *PANS*, **16**, 606-624.

Odiyo, P. O. (1973). Some observations on butterfly and dragonfly migration in the Kenya highlands, *Entomol. Mon. Mag.*, **109**, 141-147.

Odiyo, P. O. (1975). Seasonal distribution and migrations of *Agrotis ipsilon* (Lep.: Noctuidae), *Trop. Pest Bull. No. 4.* (Centre for Overseas Pest Research, London.)

Odiyo, P. O. (1979). Forecasting infestations of a migrant pest: the African armyworm *Spodoptera exempta*, *Phil. Trans. Roy. Soc. Lond.*, **B287**, 403-413.

Oke, T. R. (1978). *Boundary layer climates*, Methuen, London.

Oku, T., Chiba, T., Toki, A. & Kobayashi, T. (1976). Further notes on the early summer outbreak of the oriental armyworm on grasslands of Tohoku district, 1971, *J. Jap. Grassland Sci.*, **22**, 206-210.

Oku, T. & Kobayashi, T. (1974). Early summer outbreaks of the oriental armyworm, *Mythimna separata*, in the Tohuku district and possible causative factors, *Appl. Entomol. Zool.*, **9**, 238-246.

Oku, T. & Kobayashi, T. (1977). The oriental armyworm outbreaks in Tohoku district 1960, with special reference to the possibility of mass immigration from China, *Tohoku Nat. Agric. Expl. Stn.*, Bull. No. 55.

Oku, T. & Koyama, J. (1976). Long-range migration as a possible factor causing the late summer outbreak of the oriental armyworm, *Mythimna separata*, in Tohoku district, *Jap. J. appl. Entomol. Zool.*, **20**, 184-190.

Okubo, A., Bray, D. J. & Chiang, H. C. (1981). Use of shadows for studying the three-dimensional structure of insect swarms, *Ann. ent. Soc. Amer.*, **74**, 48-50.

Okubo, A., Chiang, H. C. & Ebbesmeyer, C. C. (1977). Acceleration field of individual midges, *Anarete pritchardi* (Dip.: Cecidomyiidae), within a swarm, *Can. Entomol.*, **109**, 149-156.

Oldroyd, H. (1964). *The natural history of flies,* Norton, New York.

Oliver, H. R. (1973). Smoke trails in a pine forest, *Weather,* **28**, 345-347.

Oliver, H. R. (1975). Wind speeds within the trunk space of a pine forest, *Quart. J. Roy. meteorol. Soc.,* **101**, 167-168.

Olivier, D. L. (1978). *Retiarius* gen. nov.: phyllosphere fungi which capture windborne pollen grains, *Trans. Br. mycol. Soc.,* **71**, 193-201.

Omer, S. M. (1979). Responses of females of *Anopheles arabiensis* and *Culex pipiens fatigens* to air currents, carbon dioxide and human hands in a flight-tunnel, *Entomol. expl. & appl.,* **26**, 142-151.

Orr, C. C. & Newton, O. H. (1971). Distribution of nematodes by wind, *Plant Dis. Repr.,* **55**, 61-63.

Orr, G. F. & Tippets, W. C, (1972). Morphology and other physical characteristics of urediospores possibly related to aerodynamics and long-range travel, *Mycopathol. & Mycol.,* **48**, 143-159.

Otis, G. W., Winston, M. L. & Taylor, O. R. (1981). Engorgement and dispersal of Africanized honeybee swarms, *J. apic. Res.,* **20**, 3-12.

Pack, D. H., Ferber, G. J., Hefter, J. L., Telegadas, K., Angell, J. K., Hoecker, W. H. & Machta, L. (1978). Meteorology of long-range transport, *Atmos. Env.,* **12**, 426-444.

Pady, S. M. (1955). The occurrence of the vector of wheat streak mosaic, *Aceria tulipae,* on slides exposed to the air, *Plant Dis. Repr.,* **39**, 296-297.

Pady, S. M. & Kelly, C. D. (1953). Studies on micro-organisms in arctic air during 1949 and 1950, *Can. J. Bot.,* **31**, 107-122.

Pasquill, F. (1974). *Atmospheric diffusion,* Ellis Horwood, Chichester, England.

Paterson, M. P. (1975). Les mouches de Point Sublime, *Weather,* **30**, 301-303.

Pedgley, D. E. (1979). *Mountain weather,* Cicerone Press, Milnthorpe, England.

Pedgley, D. E. (Ed.) (1981). *Desert locust forecasting manual, Vols. I & II,* Centre for Overseas Pest Research, London.

Pedgley, D. E. & Betts, E. (1980). Forecasting the spread of migrant insect pests. *EPPO Bull.,* **10**, 151-160.

Pérombelon, M. C. M. & Kelman, A. (1980). Ecology of the soft rot erwinias, *Ann. Rev. Phytopathol.,* **18**, 361-387.

Persson, B. (1976). Influence of weather and nocturnal illumination on the activity and abundance of populations of Noctuids (Lep.) in south coastal Queensland, *Bull. ent. Res.,* **66**, 33-63.

Pieters, E. P. & Urban, T. C. (1977). Dispersal of the boll weevil in the coastal bend area of Texas, *Southwestern Entomol.,* **2**, 4-7.

Pijl, L. van der (1972). *Principles of dispersion in higher plants.* Springer Verlag, Berlin.

Pimental, D. (Ed.) (1975). *Insects, science and society,* Academic, New York.

Polunin, N. (1954). Progress in arctic aeropalynology, *Proc. 8th Cong. Internat. Bot., Paris,* 279-281.

Polunin, N. (1955). Arctic aeropalynology. Spores observed on sticky slides exposed in various regions in 1950, *Can. J. Bot.,* **33,** 401-415.

Primault, B. (1974). La propagation d'une épizootie de fièvre aphteuse dépend-elle des conditions météorologique?, *Schweiz Arch. Tierheilk.,* **116,** 7-19.

Proctor, M. & Yeo, P. (1973). *The Pollination of flowers,* Collins, London.

Pruess, K. P. (1967). Migration of the army cutworm *Chorizagrotis auxiliaris* (Lep.: Noctuidae). I. Evidence for migration, *Ann. ent. Soc. Amer.,* **60,** 910-920.

Pulliainen, E. (1964). Studies on the humidity and light orientation and the flying activity of *Myrrha 18-guttata* (Col.: Coccinellidae), *Ann. entomol. Fenn.,* **30,** 117-141.

Rabkin, F. B. & Lejeune, R. R. (1954). Some aspects of the biology and dispersal of the pine tortoise scale, *Toumeyella numismaticum* (Hom.: Coccidae), *Can. Entomol.,* **86,** 570-575.

Ramage, C. S. (1971). *Monsoon meteorology,* Academic, New York.

Rainey, R. C. (1963). Meteorology and the migration of desert locusts, *Wld. Met. Org., Tech. Note No. 54.*

Rainey, R. C. (1973). Airborne pests and the atmospheric environment. *Weather,* **28,** 224-239.

Rainey, R. C. (1976). Flight behaviour and features of the atmospheric environment, In (Ed.) Rainey, R. C. *Insect flight,* 7th symposium, Roy. ent. Soc., London, 75-112.

Ramchandra Rao, Y. (1960). *The desert locust in India,* Indian Council of Agric. Res., Monograph No. 21.

Ramos, D. E., Moller, W. J. & English, H. (1975). Production and dispersal of ascospores of *Eutypa armeniacae* in California, *Phytopathology,* **65,** 1364-1371.

Rapilly, F. (1979). Yellow rust epidemiology, *Ann. Rev. Phytopathology,* **17,** 59-73.

Raulston, J. R. (1979). *Heliothis virescens* migration. In (Eds.) Rabb, R. L. & Kennedy, G. G. *Movement of highly mobile insects,* N. Carolina State Univ., 412-419.

Rautapää, J. (1979). Humidity reactions of cereal aphids (Hom.: Aphididae), *Ann. ent. Soc. Amer.,* **45,** 33-41.

Raygor, S. C. & Mackay, K. P. (1975). Bacterial air pollution from an activated sludge tank, *Water, Air & Soil Polln.,* **5,** 47-52.

Raynor, G. S., Cohen, L. A., Hayes, J. V. & Ogden, E. C. (1966). Dyed pollen grains and spores as tracers in dispersion and deposition studies, *J. appl. Meteorol.*, **5**, 728-729.

Raynor, G. S., Hayes, J. V. & Ogden, E. C. (1974a). Mesoscale transport and dispersion of airborne pollens, *J. appl. Meteorol.*, **13**, 87-95.

Raynor, G. S., Hayes, J. V. & Ogden, E. C. (1974b). Particulate dispersion into and within a forest, *Boundary Layer Meteorol.*, **7**, 429-456.

Raynor, G. S., Hayes, J. V. & Ogden, E. C. (1975). Particulate dispersion from sources within a forest, *Boundary Layer Meteorol.*, **9**, 257-277.

Raynor, G. S., Ogden, E. C. & Hayes, J. V. (1971). Dispersion and deposition of timothy pollen from experimental sources, *Agric. Meteorol.*, **9**, 347-366.

Raynor, G. S., Ogden, E. C. & Hayes, J. V. (1973). Dispersion of pollens from low-level crosswind line sources, *Agric. Meteorol.*, **11**, 177-195.

Raynor, G. S., Ogden, E. C. & Hayes, J. V. (1974). Enhancement of particulate concentrations downwind of vegetation barriers, *Agric. Meteorol.*, **13**, 181-188.

Reid, D. G., Wardhaugh, K. G. & Roffey, J. (1979). Radar studies of insect flight at Benalla, in February 1974, *Australia, CSIRO, Div. Entomol., Tech. Rep. No. 16.*

Reynolds, D. R. & Riley, J. R. (1979). Radar observations of concentration of insects above a river in Mali, West Africa, *Ecol. Entomol.*, **4**, 161-174.

Richardson, R. H. & Johnston, J. S. (1975). Behavioural components of dispersal in *Drosophila mimica, Oecologia*, **20**, 287-299.

Richter, C. J. J. (1967). Aeronautic behaviour in the genus *Pardosa* (Araneae: Lycosidae), *Entomol. Mon. Mag.*, **103**, 73-74.

Richter, C. J. J. (1970). Aerial dispersion in relation to habitat in eight wolf spider species, *Oecologia*, **5**, 200-214.

Ridgway, R. L., Bariola, L. A. & Hardee, D. D. (1971). Seasonal movement of boll weevils near the High Plains of Texas, *J. econ. Entomol.*, **64**, 14-19.

Riehl, H. (1979). *Climate and weather in the tropics,* Academic, New York.

Riley, D. & Spolton, L. (1974). *World weather and climate,* Cambridge Univ. Press, London.

Riley, J. R. (1975). Collective orientation in night-flying insects, *Nature*, **253**, 113-114.

Riley, J. R. (1978). Quantitative analysis of radar returns from insects. In (Eds.) Vaugn, C. R., Wolf, W. & Klassen, W. *Radar, insect population ecology, and pest management,* NASA Conf. Publicn. No. No. 2070, 131-158.

Riley, J. R. (1979). Radar as an aid to the study of insect flight. In (Eds.) Amlaner, C. J. & Macdonald, D. W. *A handbook on biotelemetry and radio tracking,* Pergamon, Oxford.

Riley, J. R. & Reynolds, D. R. (1979). Radar-based studies of the migrating flight of grasshoppers in the middle Niger area of Mali, *Proc. Roy. Soc. Lond.*, **B204**, 67-82.

Riley, J. R., Reynolds, D. R. & Farmery, M. J. (1981). Radar observations of *Spodoptera exempta*, Kenya, March-April 1979, *Centre for Overseas Pest Research, Misc. Rep. No. 54.*

Ritchie, J. C. & Lichti-Federovich, S. (1967). Pollen dispersal phenomena in arctic-subarctic Canada, *Rev. Palaeobot. Palynol.,* **3**, 255-256.

Rivnay, E. (1961). The phenology of *Prodenia litura* in Israel with reference to its occurrence in the Near East at large, *Bull. Res. Council Israel,* **108**, 100-106.

Rivnay, E. (1962). An hypothesis on the migration of the spiny cotton bollworm, *Earias insulana* in Israel, *Proc. 11th Internat. Congr. Entomol.,* **3**, 40.

Roach, S. H. & Ray. L. (1972). Boll weevils captured at Socastee, South Carolina, in 1970, in wing traps placed around fields with and without growing cotton, *J. econ. Entomol.,* **65**, 559-560.

Rockett, T. R. & Kramer, C. L. (1974). The biology of sporulation of selected Tremellales, *Mycologia,* **66**, 926-941.

Roelofs, W. L. (1978). Threshold hypothesis for pheromone perception, *J. Chem. Ecol.,* **4**, 685-699.

Roffey, J. (1963). Observations on night flight in the desert locust, *Anti-Locust Bull. No. 39.* (Centre for Overseas Pest Research, London.)

Rose, D. J. W. (1979). The significance of low-density populations of the African armyworm *Spodoptera exempta, Phil. Trans. Roy. Soc. Lond.,* **B287**, 392-402.

Rose, D. J. W. & Dewhurst, C. F. (1979). The African armyworm, *Spodoptera exempta* — congregation of moths in trees before flight, *Entomol. exp. & appl.,* **26**, 346-348.

Rowe, R. C. & Beute, M. K. (1975). Ascospore formation and discharge by *Calonectria crotalariae, Phytopathology,* **65**, 393-398.

Rummel, D. R., Jordan, L. B., White, J. R. & Wade, L. J. (1977). Seasonal variation in the height of boll weevil flight, *Env. Entomol.,* **6**, 674-678.

Safranyik, L. (1978). Effects of climate and weather on mountain pine beetle populations. In (Eds.) Berryman, A. A., Amman, G. D., Stark, R. W. & Kibbee, D. L. *Theory and practice of mountain pine beetle management in lodgepole pine forests,* Moscow, Univ. Idaho For., Wildl., Range Exp. Stn., 77-84.

Salama, H. S. & Shoukry, A. (1972). Flight range of the moth of the cotton leafworm *Spodoptera littoralis, Z. ang. Entomol.,* **71**, 181-184.

Saxena, R. C. & Justo, H. D. (1980). *Long distance migration of brown planthopper in the Philippine archipelago.* (Unpublished paper presented to the 11th Annual Conference of the Pest Control Council of the Philippines.)

Schaefer, G. W. (1970). Radar studies of locust, moth and butterfly migrations in the Sahara, *Proc. Roy. ent. Soc.,* **C34**, 3, 39-40.

Schaefer, G. W. (1972a). Radar studies of the flight activity and aerial transportation of insects, *Proc. 14th Internat. Cong. Entomol., Canberra,* 154-155.

Schaefer, G. W. (1972b). Radar detection of individual locusts and swarms, *Proc. Internat. Study Conf. Current and Future Problems of Acridology, London 1970,* 379-380.

Schaefer, G. W. (1976). Radar observation of insect flight. In (Ed.) Rainey, R. C. *Insect flight,* 7th symposium, Roy. ent. Soc., Lond., 157-197.

Schieber, E. (1975). Present status of coffee rust in South America, *Ann. Rev. Phytopathology,* **13**, 375-382.

Schmidt-Koenig, K. (1978). Directions of migrating monarch butterflies (*Danaus plexippus*) in some parts of the eastern United States, *Behavioural Processes,* **4**, 73-78.

Schroth, M. N., Thomson, S. V., Hildebrand, D. C. & Moller, W. J. (1974). Epidemiology and control of fireblight, *Ann. Rev. Phytopathology,* **12**, 389-412.

Scorer, R. S. (1978). *Environmental aerodynamics,* Ellis Horwood, Chichester.

Scott, P. R. & Bainbridge, A. (1978). *Plant disease epidemiology,* Blackwell, Oxford.

Scriven, R. A. & Fisher, B. E. A. (1975). The long range transport of airborne material and its removal by deposition and washout, *Atmos. Env.,* **9**, 49-58, 59-68.

Sellers, R. F. (1980). Weather, host and vector — their interplay in the spread of insect-borne animal virus diseases, *J. Hyg., Camb.,* **85**, 65-102.

Sellers, R. F. & Forman, A. J. (1973). The Hampshire epidemic of foot-and-mouth disease, 1967, *J. Hyg., Camb.,* **71**, 15-34.

Sellers, R. F., Gibbs, E. P. J., Herniman, K. A. J., Pedgley, D. E. & Tucker, M. R. (1979). Possible origin og the bluetongue epidemic in Cyprus, August 1977. *J. Hyg., Camb.,* **83**, 547-555.

Sellers, R. F. & Parker, J. (1969). Airborne excretion of foot-and-mouth disease virus, *J. Hyg., Camb.,* **67**, 671-677.

Sellers, R. F., Pedgley, D. E. & Tucker, M. R. (1977). Possible spread of African horse sickness on the wind, *J. Hyg., Camb.,* **79**, 279-298.

Sellers, R. F., Pedgley, D. E. & Tucker, M. R. (1978). Possible windborne spread of bluetongue to Portugal, June–July 1956, *J. Hyg., Camb.,* **81**, 189-196.

Shaw, M. W. (1962). The diamond-back moth migration of 1958, *Weather,* **17**, 221-234.

Shorey, H. H. & McKelvey, J. J. (1977) (Eds.). *Chemical control of insect behaviour,* Wiley.

Slykhuis, J. T. (1955). *Aceria tulipae* (Acarina: Eriophyidae) in relation to spread of wheat streak mosaic, *Phytopathology,* **45**, 116-128.

Smith, C. V. (1964). Some eividence for the windborne spread of fowl pest, *Meteorol. Mag.*, **93**, 257-263.

Smith, L. P. (1970). *Weather and animal diseases, Wld. Met. Org., Tech. Note 113.*

Snow, W. F. (1979). The vertical distribution of flying mosquitoes (Dip.: Culicidae) near an area of irrigated rice-fields in the Gambia, *Bull. ent. Res.*, **69**, 561-571.

Solbreck, C. (1980). Dispersal distances of migrating pine weevils, *Hylobis abietis* (Col.: Curculionidae), *Entomol. exp. & appl.*, **28**, 123-131.

Sotthibandhu, S. & Baker, R. R. (1979). Celestial orientation by the large yellow underwing moth, *Nactua pronuba, Anim. Behav.*, **27**, 786-800.

Sower, L. L., Kaae, R. S. & Shorey, H. H. (1973). Sex pheromones of Lepidoptera. XLI. Factors limiting potential distance of sex pheromone communication in *Trichoplusia ni, Ann. ent. Soc. Amer.*, **66**, 1121-1122.

Sparks, A. N., Jackson, R. D. & Allen, C. L. (1975). Corn earworms: capture of adults in light traps on unmanned oil platforms in the Gulf of Mexico, *J. econ. Entomol.*, **68**, 431-432.

Spieksma, F. T. M. (1980). Daily hay-fever forecast. In (Ed.) Federal Environmental Agency (of West Germany), *Proc. 1st Internat. Conf. Aerobiol., Munich 1978*, 307-315.

Spotts, R. A., Stang, E. J. & Ferree, D. C. (1976). Effect of overtree misting for bloom delay on incidence of fireblight in apple, *Plant Dis. Repr.*, **60**, 329-330.

Stadelbacher, E. A. & Ffrimmer, T. R. (1972). Winter survival of the bollworm at Stoneville, Mississippi, *J. econ. Entomol.*, **65**, 1030-1034.

Starr, J. R. (1967). Deposition of particulate matter by hydrometeors, *Quart. J. Roy. meteorol. Soc.*, **93**, 516-521.

Starr, J. R. & Mason, B. J. (1966). The capture of airborne particles by water drops and simulated snow crystals, *Quart. J. Roy. meteorol. Soc.*, **92**, 490-499.

Stedman, O. J. (1979). Patterns of unobstructed splash dispersal, *Ann. appl. Biol.*, **91**, 271-285.

Stedman, O. J. (1980a). Splash droplet and spore dispersal studies in field beans (*Vicia faba*), *Agric. Meteorol.*, **21**, 111-127.

Stedman, O. J. (1980b). Splash dispersal studies in wheat using a fluorescent tracer, *Agric. Meteorol.*, **21**, 195-203.

Stedman, O. J. (1980c). Observations on the production and dispersal of spores, and infection by *Rhynchosporium secalis, Ann. appl. Biol.*, **95**, 163-175.

Stern, V. M. (1979). Long and short range dispersal of the pink bollworm *Pectinorphora gossypiella* over southern California, *Env. Entomol.*, **8**, 524-527.

Stern, V. & Sechaverian, V. (1978). Long-range dispersal of pink bollworm into the San Joaquin Valley, *Calif. Agric.*, **32**, 4-5.

Stephens, G. R. & Aylor, D. E. (1978). Aerial dispersal of red pine scale, *Matsucoccus resinosae* (Hom.: Margarodidae), *Env. Entomol.*, **7**, 556-563.

Sullivan, R. T. (1981). Insect swarming and mating, *Fla. Entomol.*, **64**, 44-65.

Suomalainen, E. & Mikkola, K. (1967). *Nycteola asiatica* (Lep.: Noctuidae) as a migrant in northern Europe, *Ann. entomol. Fenn.*, **33**, 102-107.

Sutton, J. C., Swainton, C. J. and Gillespie, T. J. (1978). Relation of weather variables and host factors to incidence of airborne spores of *Botrytis squamosa, Can. J. Bot.*, **56**, 2460-2469.

Sutton, T. B., Jones, A. L. & Nelson, L. A. (1976). Factors affecting dispersal of conidia of the apple scab fungus, *Phytopathology*, **66**, 1313-1317.

Suzuki, H., Hayashi, K. & Asahina, S. (1977). Note on the transoceanic insects captured on East China Sea in 1976, *Trop. Med.*, **19**, 85-93.

Sylvén, E. (1970). Field movement of radioactively labelled adults of *Dasyneura brassicae* (Dip.: Cecidomyiidae), *Entomol. Scand.*, **1**, 161-187.

Symmons, P. M. (1978). The prevention of plagues of the red locust, *Nomadacris septemfasciata, Acrida*, **7**, 55-78.

Symmons, P. M. & Luard, E. (1980). The simulated distribution of night flying insects in a wind convergence, *Australian Plague Locust Commission, Ann. Rep. 1978-9, Res. Suppl.*, 20-36.

Symmons, P. M. & McCulloch, L. (1980). Persistence and migration of *Chortoicetes terminifera* in Australia, *Bull. ent. Res.*, **70**, 197-201.

Taimr, L. & Kriz, J. (1978). Stratiform drift of the hop aphid (*Phorodon humuli*), *Z. ang. Entomol.*, **86**, 71-79.

Taimr, L., Kudlelová, A. & Kriz, J. (1978). Diurnal periodicity in the flight activity of migrant alatae of *Phorodon humuli, Z. ang. Entomol.*, **86**, 373-380.

Tampieri, F., Mandrioli, P. & Puppi, G. L. (1977). Medium range transport of airborne pollen, *Agric. Meteorol.*, **18**, 9-20.

Taylor, C. E. & Johnson, C. G. (1954). Wind direction and the infestation of beanfields by *Aphis fabae, Ann. appl. Biol.*, **41**, 107-116.

Taylor, L. R. (1958). Aphid dispersal and diurnal periodicity, *Proc. Linn. Soc. Lond.*, **169**, 67-73.

Taylor, L. R. (1960). The distribution of insects at low levels in the air, *J. anim. Ecol.*, **29**, 45-63.

Taylor, L. R. (1973). Monitor survey for migrant insect pests, *Outlook Agric.*, **7**, 109-116.

Taylor, L. R. (1974). Insect migration, flight periodicity and the boundary layer, *J. anim. Ecol.*, **43**, 225-238.

Taylor, L. R. (1977). Migration and the spatial dynamics of an aphid. *Myzus persicae, J. anim. Ecol.*, **46**, 411-423.

Taylor, L. R., Brown, E. S. & Littlewood, S. C. (1979). The effect of size on the height of flight of migrant moths, *Bull ent. Res.*, **69**, 605-609.

Taylor, L. R., French, R. A. & Macaulay, E. D. M. (1973). Low-altitude migration and diurnal flight periodicity; the importance of *Plusia gamma* (Lep.: Plusiidae), *J. anim. Ecol.*, **42**, 751-760.

Taylor, L. R. & Taylor, R. A. J. (1977). Aggregation, migration and population mechanics, *Nature*, **265**, 415-421.

Taylor, L. R., Woiwod, I. P. & Taylor, R. A. J. (1979). The migratory ambit of the hop aphid and its significance in aphid population dynamics, *J. anim. Ecol.*, **48**, 955-972.

Taylor, O. R. (1977). The past and possible future spread of Africanised honeybees in the Americas, *Bee World*, **58**, 19-30.

Taylor, R. A. J. (1978). The relationship between density and distance of dispersing insects, *Ecol. Entomol.*, **3**, 63-70.

Teltsch, B. & Katzenelson, E. (1978). Airborne enteric bacteria and viruses from spray irrigation with wastewater, *Appl. env. Microbiol.*, **35**, 290-296.

Thomas, G. & Wood, F. (1980). Colorado beetle in the Channel Islands, *EPPO Bull.*, **10**, 491-498.

Thresh, J. M. (1966). Field experiments on the spread of black current reversion virus and its gall mite vector (*Phytoptus ribis*), *Ann. appl. Biol.*, **58**, 219-230.

Thresh, J. M. (1976). Gradients of plant virus diseases, *Ann. appl. Biol.*, **82**, 381-406.

Thresh, J. M. (1981) (Ed.). *Pests, pathogens and vegetation*, Pitman, London.

Thygeson, T. (1968). Insect migration over long distances, *Saetr. ugeskr. Agron.*, **8**, 115-120.

Tinline, B. (1970). Lee wave hypothesis for the initial pattern of spread during the 1967-68 foot-and-mouth epizootic, *Nature*, **227**, 860-862.

Tomlinson, A. I. (1973). Meteorological aspects of trans-Tasman insect dispersal, *New Zea. Entomol.*, **5**, 253-268.

Traynier, R. M. M. (1968). Sex attraction in the Mediterranean flour moth, *Anagasta kuhniella:* location of the female by the male, *Can. Entomol.*, **100**, 5-10.

Turnock, W. J., Gerber, G. H. & Bickis, M. (1979). The applicability of X-ray energy-dispersive spectroscopy to the identification of populations of red turnip beetle, *Entomoscelis americana* (Col.: Chrysomelidae), *Can. Entomol.*, **111**, 113-125.

Tyldesley, J. B. (1973). Long-range transmission of tree pollen to Shetland — I. Sampling and trajectories, *New Phytol.*, **72**, 175-181.

Urquhart, F. A. (1960). *The monarch butterfly*, Univ. Toronto Press.

Urquhart, F. A. & Urquhart, N. R. (1977). Overwintering areas and migratory routes of the monarch butterfly (*Danaus plexippus*) in North America, with special reference to the western population, *Can. Entomol.*, **109**, 1583-1589.

Urquhart, F. A. & Urquhart, N. R. (1979). Aberrant autumnal migration of the eastern population of the monarch butterfly, *Danaus plexippus,* as it relates to the occurrence of strong westerly winds, *Can. Entomol.,* **111,** 1281-1286.

Vale, G. A. (1977). The flight of tsetse flies (Dip.: Glossinidae) to and from a stationary ox, *Bull. ent. Res.,* **67,** 297-303.

Vale, G. A. (1980). Flight as a factor in the host-finding of tsetse flies, *Bull. ent Res.,* **70,** 299-307.

Venette, J. & Kennedy, B. W. (1975). Naturally produced aerosols of *Pseudomonas glycinea, Phytopathology,* **65,** 737-738.

Vugts, H. F. & Wingerden, W. K. R. E. van (1976). Meteorological aspects of aeronautic behaviour of spiders, *Oikos,* **27,** 433-444.

Wainhouse, D. (1979). Dispersal of the beech scale (*Cryptococcus fagi*) in relation to the development of beech bark disease, *Mitt. Schweiz. entomol. Ges.,* **52,** 181-183.

Wainhouse, D. (1980). Dispersal of first instar larvae of the felted beech scale, *Cryptococcus fagisuga, J. appl. Ecol.,* **17,** 523-532.

Walker, T. J. & Riordan, A. J. (1981). Butterfly migration: are synoptic-scale wind systems important? *Ecol. Entomol.,* **6,** 433-440.

Wall, C. & Perry, J. N. (1980). Effects of spacing and trap number on interactions between pea moth pheromone traps, *Entomol. exp. & appl.,* **28,** 313-321.

Waller, J. M. (1972). Water-borne spore dispersal in coffee berry disease and its relation to control, *Ann. appl. Biol.,* **71,** 1-18.

Waller, J. M. (1979). The recent spread of coffee rust (*Hemileia vastatrix*) and attempts to control it. In (Eds.) Ebbels, D. L. & King, J. E. *Plant health,* 275-283.

Waller, J. M. (1981). The recent spread of some tropical plant diseases, *Trop. Pest Management,* **27,** 360-362.

Waloff, Z. (1963). Field studies in solitary and *transiens* desert locusts in the Red Sea area, *Anti-locust Bull. No. 40.* (Centre for Overseas Pest Research, London.)

Waloff, Z. (1966). The upsurges and recessions of the desert locust plague: an historical survey, *Anti-locust Memoir No. 8.* (Centre for Overseas Pest Research, London.)

Waloff, Z. (1972). Orientation of flying locusts, *Schistocerca gregaria,* in migrating swarms, *Bull. ent Res.,* **62,** 1-72.

Waloff, Z. & Green, S. (1975). Regularities in duration of regional desert locust plague, *Nature,* **256,** 484-485.

Walsh, J. F., Davies, J. B. & Garms, R. (1981). Further studies on the reinvasion of the onchocerciasis control programme by *Simulium damnosum,* s. l.: the effects of an extension of control activities into southern Ivory Coast during 1979, *Tropenmed. Parasit.,* **32,** 269-273.

Wang, C-W., Perry, T. O. & Johnson, A. G. (1960). Pollen disperion of slash pine (*Pinus elliottii*) with special reference to seed orchard management, *Silvae genet.*, **9**, 78-86.

Wang, Y-C. (1980). A study of the synchronisation of the occurrence of two noctuids: the black cutworm and the armyworm, *Acta Phytophylatica Sin.*, **7**, 247-251.

Way, M. J., Cammell, M. E., Taylor, L. R. and Woiwod, I. P. (1981). The use of egg counts and suction trap examples to forecast the infestation of spring-sown field beans, *Vicia faba*, by the black bean aphid, *Aphis fabae*, *Ann. appl. Biol.*, **98**, 21-34.

Wellman, F. L. & Echandi, E. (1981). The coffee rust situation in Latin America in 1980, *Phytopathology*, **71**, 968-970.

Wilkinson, A. G. & Spiers, A. G. (1976). Introduction of the poplar rusts *Melampsora larici-populina* and *M. medusae* to New Zealand and their subsequent distribution, *New Zea. J. Sci.*, **19**, 195-198.

Willard, J. R. (1973). Survival of crawlers of California red scale *Aonidiella aurantii* (Hom.: Diaspididae), *Australian J. Zool.*, **21**, 567-573.

Willard, J. R. (1974). Horizonatal and vertical dispersal of California red scale, *Aonidiella aurantii* in the field, *Australian J. Zool.*, **22**, 531-548.

Willard, J. R. (1976). Dispersal of California red scale *Aonidiella aurantii* in relation to weather variables, *J. Australian entomol. Soc.*, **15**, 395-404.

Williams, C. B., Wenz, J. M., Dahlsten, D. L. & Norick, N. X. (1979). Relation of forest site and stand characteristics to Douglas-fir tussock moth outbreaks in California, *Mitt. Schweiz ent. Ges.*, **52**, 297-307.

Willis, H. R. (1939). Painting for determination of grasshopper flights, *J. econ. Entomol.*, **32**, 401-403.

Wingerden, W. K. R. E. van & Vugts, H. F. (1980). Ecological and meteorological aspects of aeronautic dispersal of spiders, In (Ed.) Federal Environmental Agency (of West Germany), *Proc. 1st Internat. Conf. Aerobiol., Munich 1978*, 212-219.

Winktelius, S. (1977). The importance of southerly winds and other weather data on the incidence of sugar beet yellowing viruses in southern Sweden, *Swedish J. agric. Res.*, **7**, 89-95.

Winktelius, S. (1980). Aerial dispersal of aphids into Sweden. In (Ed.) Federal Environmental Agency (of West Germany), *Proc. 1st Internat. Conf. Aerobiol., Munich 1978*, 220-226.

Wright, D. M., Bailey, G. D. * Hatch, M. T. (1968). Survival of airborne mycoplasma as affected by relative humidity, *J. Bacteriol.*, **95**, 251-252.

Wright, J. W. (1953). Pollen-dispersion studies: some practical applications, *J. For.*, **51**, 114-118.

Yoshino, M. M. (1975). *Climate in a small area*, Univ. Tokyo Press.

Young, J. R. (1979). Assessing the movement of the fall armyworm (*Spodoptera frugiperda*) using insecticide resistance and wind patterns. In (Eds.) Rabb, R. L. & Kennedy, G. G. *Movement of highly mobile insects*, N. Carolina State Univ., 344-351.

Index